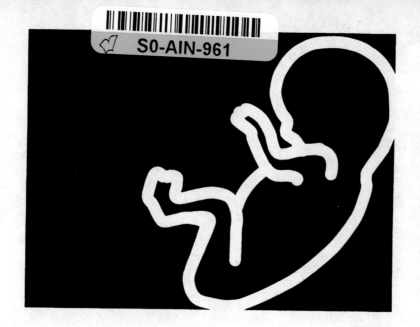

Conception to Birth

Howard M. Lenhoff

KENDALL/HUNT PUBLISHING COMPANY
4050 Westmark Drive Dubuque, Iowa 52002

CONCEPTION TO BIRTH
Human Reproduction, Genetics, and Development

Howard M. Lenhoff

Professor of Developmental and Cell Biology
University of California, Irvine

KENDALL/HUNT PUBLISHING COMPANY
4050 Westmark Drive   Dubuque, Iowa 52002

DEDICATION

To my daughter, Gloria,
the inspiration for this text
and an inspiration to those
whose paths she has crossed.

PROLOGUE

Perhaps the most beautiful and complex series of events that takes place on this earth is triggered the moment that the human egg and sperm join. The eventual product of this union, if all goes well, is the creation of a human life. In one sense, then, most of us are endowed with an extraordinary power of being able to participate in the creation of human life. At the moment that we participate in this act of creation, of love, however, we not only assume a horrendous responsibility, but our entire lives become unalterably changed — for better or for worse.

Whether for better or for worse may depend on our genetic makeup and on our behavior and activities before the union takes place, and especially during the nine months of gestation. Hence, the object of this book: to provide you with fundamental information on the biology of human reproduction, genetics, and development so that you can make intelligent decisions, decisions that will do much to assure that you will have healthy children who are as far as possible free of both mental and physical defects.

What are those critical biological events preceding and following conception? Aren't they out of our control once fertilization commences? Do we really need to understand them? After all, didn't our parents and their parents and so on produce healthy babies before the current explosion of knowledge about the biology of reproduction?

True - partly, that is. Most of the children who survived did well. But what of those who did not survive, who died in infancy, who were hidden from view, who "could not compete," or who in some cases in the not too distant past were burned as witches because of their handicaps?

Today we do live in a different world. On the negative side, we are exposed to more factors, such as food additives, drugs, radiation, and toxic substances, which may increase the probability of our having children with some type of physical or mental handicap. On the positive side, our new advances in the biological and medical sciences give us new information, medicines, and procedures which can greatly enhance the odds for having healthy children, and, for some people who in earlier times would have been childless, of conceiving children.

This book, therefore, is written to provide you in a simple, clear and, I hope, interesting manner with the basic principles of biology that will allow you to make life-shaping decisions in those critical months before and after fertilization. I stress basic principles and not facts, because some of our "facts" are still changing and, through research, we are uncovering new facts almost daily. On the other hand, if you have an understanding of the principles, if you critically evaluate the facts on hand at the time you are thinking of having children, then your chances of being a successful creator will be greatly increased. You could take your chances . . . but the simple fact that you are reading this book indicates that you do care.

The book is organized into three parts: human reproduction, human genetics, and human embryonic and fetal development through birth. The first part deals with the male and female reproductive systems, the role of hormones in regulating the production and release of eggs and sperm, the female menstrual cycle, the sexual response, and sexually transmitted diseases including AIDS. You will note that there is no special chapter dealing with the practicalities of contraceptives. Instead, contraceptives are dealt with at a number of places throughout the book, at points where the relevant biological processes basic to contraception are first explained. For example, it would be useless to try to explain steroid oral contraceptive pills without first providing an understanding of how steroids affect the release of eggs from the ovary.

The second part of this book deals with a number of aspects of human genetics. It begins with a description of the processes by which cells multiply and distribute the genetic material into the egg and sperm. This section is followed by a description of a major category of birth defects which can result from biological "mistakes" made while eggs and sperm form and unite. Next the basic principles of genetics are discussed, that is the processes by which the genes are distributed to the baby and are expressed. This section ends with a discussion of the molecular aspects of genetics in order that we may understand mutation, genetic diseases, and modern techniques in gene biotechnology.

The third and final part of the book deals with fertilization, the placement of the early embryo into the wall of the uterus, and the development of the embryo, with special emphasis placed on the origin of the brain, nervous system, digestive system, heart, and the reproductive organs. It tells of the formation and role of the placenta, the effect of pregnancy on the physiology of the mother, the role of such environmental factors as drugs, viruses, and radiation in causing birth defects, and the development of the fetus. The book ends with a discussion of the physiology of birth, and the process by which the mother makes and delivers milk to the newborn infant.

At the appropriate points in the book, some of the newer technologies of "test tube babies," gene transfer, and amniocentesis are discussed. On the other hand, this book will not deal with those psychological, sociological, and educational factors that are equally important and that also will affect the well-being of the children you bear.

My intent is to give you a basic understanding of the biological processes your body goes through while it is creating a human being. I hope that you will keep the book and reread it when you are planning to have children. The insights you gain from these pages should help you to comprehend the new medical information regarding human reproduction, genetics and embryology that appears from time to time in the better newspapers and magazines.

You will note that I try to use simple English words instead of Greek and Latin terms to describe some of the biological structures and processes described in this book. For those of you who wish to read more advanced scientific literature in this field, however, I have placed the technological names in brackets following my first use of the simplified English version. My primary goal is not to provide you with the technical vocabulary of reproductive biology, but rather to give you an understanding of your own reproductive processes so that you will be a knowledgeable creator of human life.

Finally, I make a special attempt to include in this book only the material that I believe absolutely necessary for you to understand human reproduction, human genetics, and human embryonic development, and I purposely leave out much information and many more terms that I could have included. Thus, if you want more information, I suggest you go to the library and look at some of the more advanced articles and books in the field. I think you will find that after studying the material in this book you will be adequately prepared to delve into such advanced readings.

ACKNOWLEDGEMENTS

This text was written for a course that I devised at the University of California, Irvine, for students who were not biology majors. The course aimed to provide them with the fundamental facts and principles of that most precious of God's gifts, the gift of procreation. For the first ten years I taught this course, I prepared for those students a complete set of newly revised lecture notes, and every year my students offered further suggestions of ways to improve those notes. The contents of this book, then, are the results of the evolution of those notes in response to the constructive comments of literally thousands of students. I take great pleasure in thanking them for guiding me in writing a book to fulfill the sole purpose of serving students who want to understand the biology of "conception to birth."

In preparing this book, I had the help of numerous other people. For detailed editing and criticism, I thank my wife and scholarly collaborator in our research on the history of science, Sylvia G. Lenhoff. For additional editorial comments, I thank my departmental colleague, Dr. Ronald Meyer, who teaches his own special version of a course similar to mine. And I thank my son, Bernie Lenhoff, now M.F.A., who offered comments on the book in its earliest drafts from the perspective of an undergraduate student.

My experience in bringing this text to completion was unique, different from the methodology employed with the nine other books that I have edited or coauthored, because it was presented to the publisher camera ready. For formatting the entire text, and for preparing all of the drawings, including the original art and the computer aided ones, I thank the very talented and dedicated Ms. Leslie Wolcott. I did all of the typing.

I thank Dr. Michael Gonzalez for introducing me to Kendall/Hunt Publishing Company, and Associate Editor Janice Samuells for prodding me to finish the book on schedule.

Finally, I never would have attempted to publish such a book as this but for my daughter Gloria (see Dedication and Epilogue). I believe that after you have read and assimilated the material in *Conception to Birth*, you too will thank her.

Please save the book after you finish your course. It should serve you well as you start your family.

Howard M. Lenhoff
Professor of Biological and Social Sciences
University of California, Irvine
September 15, 1989

CONTENTS

Section III: Human Development from Fertilization through Birth

Chapter 1

An Introduction to Sexual Reproduction

Why Sex?

Most of you have a ready tongue-in-cheek answer to the question "Why sex?" But do you really understand the importance and the process of *sexual reproduction*, the process which is responsible for plant and animal life on this earth as we now know it? Every species of animal life reproduces in order to increase in number, but must that reproduction be the result of a sexual union between two partners?

Asexual Reproduction and Clones

Not necessarily. There are many examples in biology in which reproduction may take place *asexually*, that is without sex. For example, single-cell amebas reproduce by simply splitting apart (Fig. 1A). Even more complex animals, such as sea anemones, reproduce by a sort of *fission* whereby an anemone literally tears itself apart into two separate animals (Fig. 1-1B).

Some sea anemones reproduce by a more bizarre asexual process whereby an individual can convert part of its single tubular foot into a flat disc that then severs itself from the parent animal as a flat "donut," or "bagel." This flattened donut then divides into five or six pieces, each of which *regenerates* a new tiny sea anemone within several days (Fig.1-1C). If human beings reproduced in this way, it would be as if you went to the delivery room, stood in your bare feet on a sticky floor, and walked away leaving your toes behind, with each toe then developing into a baby!

Actually, the first case of asexual reproduction was discovered only about 250 years

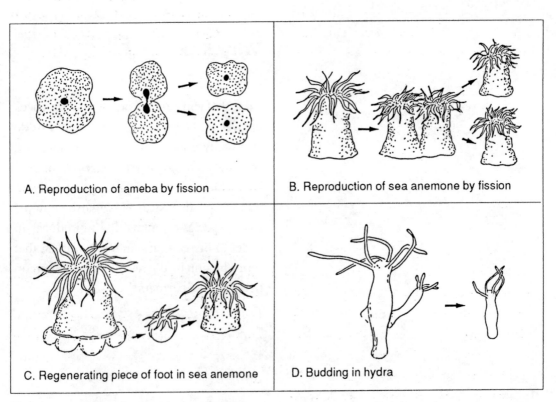

A. Reproduction of ameba by fission

B. Reproduction of sea anemone by fission

C. Regenerating piece of foot in sea anemone

D. Budding in hydra

Figure 1-1. Forms of asexual reproduction

ago in a freshwater cousin of sea anemones, the hydra. This simple tubular animal reproduces asexually by giving off "buds," almost in the same manner as do some plants. By this asexual process of budding, a hydra will grow to a certain size and then develop a small hydra along its body. In a few days that small hydra will separate from the parent, will feed by itself, and will begin reproducing itself by the same asexual process (Fig.1- 1D).

These four examples of asexual reproduction, and there are many more, all lead to the production of *clones* of individuals, i.e. individuals that are genetically alike. That means that all of the cells of all individuals in a clone have the identical number and kinds of chromosomes and genes that were present in the parent before it reproduced asexually. In these animals, as the word *asexual* implies, the reproductions described do not make use of the key cells in sexual reproduction, the egg and sperm.

Parthenogenesis

Can there be asexual reproduction that involves the egg or sperm? Yes, but this form of reproduction, called *parthenogenesis* (Greek for "to be born of a virgin") involves only the egg cells. In a number of organisms, such as plant lice (aphids) where this process was first discovered, sea urchins, some fish, and turkeys, the eggs starts to divide with-

out the involvement of sperm, and form clones of complete individuals. The chromosomes and genes in these individuals are the same ones that were present in the egg!

Human Cloning

Can parthenogenesis occur in humans? Not to our knowledge. But the word "clone" may provoke thoughts on current science fiction stories of human clones. The cloning of animals, which can be carried out with frogs, for example, is a process by which the nucleus of a cell from a donor frog is substituted for the nucleus of an egg cell. When that "egg" develops, the new frog will have the same genes as were originally present in the cells of the donor animal, whether it was male or female. Cloning actually takes the "sex" out of reproduction.

Sex, Genotype, Evolution, and Survival

All very interesting, but still we have not answered the question "Why sex?" The answer lies in the term "genetic make-up," or *genotype*. Recall that asexual reproduction and cloning lead to individuals all having identical genes. We say they are of the same *genotype*, i.e. they are identical in every way as far as their inherited characteristics are concerned.

Are such forms of reproduction desirable? They would be if we lived in a stable environment in which we had all the capacities needed to survive. But our planet has been constantly changing for billions of years. Once it was covered with water. Then bodies of land emerged. Climatic changes, such as glaciers and extremes of temperature, have prevailed at various epochs in the history of the earth. The availability of food changed with these geoclimactic periods. Any organism that was genetically fixed in its abilities to exist might not survive if it could not adapt to those changes.

The key to survival in a changing environment, then, is the ability to alter one's genetic capacities, to provide for genetic variability, or, in short, to evolve into individuals with different capacities to survive. The primary way for a species to obtain those capacities is to produce members of the species having varying genetic capacities. If even only a few members of a species possess the genes which allow them to cope with the changed environment, then they will survive, reproduce, and thereby preserve their species. This, then, is the answer to the question "Why sexual reproduction?" — to allow for genetic variability through the mixing of genes that occurs during fertilization, and, therefore, for evolution so that the species will survive.

Without sex there would not be evolution. And it is through sexual reproduction that we are doing our part, not only to reproduce more members of our species and of our particular family, but also to contribute to the survival of the human species. Even hydra and sea anemones, those animals which reproduce asexually, have a phase in their lifetimes during which they too reproduce sexually, i.e. by means of eggs and sperm, because it enhances their capacities for survival in a changing environment.

You may think that much of the foregoing is rather esoteric for the average person, if interesting to the scientist, and that it really has little immediate relevance for those of us concerned about life on this planet during the next sixty years or so. After all, there is little chance of any major geoclimactic change occurring in our lifetimes; those things take millions of years. Correct, that was true in the past when man had little effect on those changes. Today, however, through man's technological breakthroughs, we have the hydrogen bomb, breaks in the ozone layer, the greenhouse effect, acid rain, defoliation, pollution, and the mass slaughter of game animals for sport and commercial exploitation. Thus, whether or not some of the species that we now know will exist in the times of our grandchildren will depend upon how we control those man-made factors and how successfully those species are able to adapt through the genetic changes allowed for by sexual reproduction.

Overview of Some Major Topics in Sexual Reproduction

Now you have some idea of why most animal life on this earth depends upon sexual reproduction. Next, in order to give you an idea of the material to be covered in the first half of this book, I present an overview of: (1) *formation* of eggs and sperm; (2) factors controlling the maturation, *viability*, and *release* of germ cells; and (3) *transport* and union of germ cells. I will present a discussion of the establishment of *pregnancy* and maintenance of *normal development*, and of *birth* and immediate *survival* of the young in the last part of this book.

Formation of Germ Cells

There are some important terms that you need to know. One is *germ cell*. A germ cell does not refer to germs, i.e. to some sort of bacteria or virus that should be avoided to prevent illnesses. Germ cells, also called *gametes*, refer specifically to either egg cells (*ovum*, singular, or *ova*, plural) and *sperm*. All other cells in your body, such as muscle, nerve, skin, and blood cells, are called *somatic cells*. Most cells are not as large as the dot over the letter i. All human cells, germ and somatic, contain a circular body (*nucleus*) (Fig. 2-6) containing *chromosomes* (Fig. 9-1). Chromosomes are threadlike structures that contain *genes*, the material of the cell from which our inherited characteristics are derived. Genes are made up of a chemical known as DNA, which is short for deoxyribonucleic acid (Chapt. 16).

The human somatic cell contains two sets of 23 chromosomes, or 46 in all. The somatic cells of some animals may have more or less than 46 chromosomes. That number, however, will always be an even number, and will be the same for all normal members of that species.

Gametes, i.e. germ cells, however, always contain half the number of chromosomes found in somatic cells. Human germ cells, therefore, contain 23 chromosomes. On fertilization, when the sperm and egg combine, we get a fertilized egg containing 46 chromosomes, 23 contributed by the father and 23 by the mother. All of the somatic cells derived from that fertilized egg will have exact copies of those same 46 chromosomes, 23 from each parent. Thus, it is this union of human germ cells taking place during fertilization, and the accompanying mixing of the two sets of 23 chromosomes from each parent, that provides the basis of our genetic

male germ cell 23
female "   "   23
46 chromosomes

germ cells = 23 chromosome

Somatic cells = 46 chromosomes

5

variability, the best chance for survival of our species.

The processes leading to the formation of germ cells are extremely important if we are to become parents of normal healthy children, or if we are to become parents at all. People who cannot form viable germ cells, or who form abnormal germ cells may be healthy in all respects and may live to a ripe old age. Yet if they cannot form germ cells, or if they form them in insufficient numbers, these people are for all practical purposes sterile or infertile, that is they cannot have children through intercourse with the opposite sex. Through such new technologies as "test tube babies," however, some of these individuals may now conceive and bear children.

In other cases, a person may produce germ cells regularly and in sufficient amounts, but some of those germ cells may have an abnormal number of chromosomes, like 22 or 24. Children formed from the union of one of those germ cells with a normal one may suffer from such birth defects as Down's syndrome (Trisomy 21; "Mongolism"), a condition in which each of their cells has three "Number 21" chromosomes rather than two of those chromosomes as do individuals not having Down's syndrome.

There are also cases in which a gene in the germ cell may be altered and eventually expressed in a child, grandchild, or great grandchild. These altered genes, called *mutations*, can lead to one of many **genetic diseases**, such as hemophilia (inability to stop bleeding), cystic fibrosis, or Tay-Sachs disease, that may be inherited in subsequent generations. Not all mutations are harmful, however. Some mutations can lead to positive permanent changes that will allow for evolution and survival of the species.

If there is a family history of deleterious mutations, i.e. those that lead to an inherited disability, prospective parents can lessen the odds of those mutations being expressed in their children. Such parents should take advantage of genetic counseling and of prenatal counseling and testing.

Maturation, Viability, & Release of Gametes

For an egg to become fertilized, the factors, mostly neural and hormonal, controlling the *maturation, viability* (survival), and *release* of the egg and sperm must be in near perfect order. As regards the time in life germ cells mature, they must be ready during the period when the adult is prepared to care for the child. Mature germ cells are of no use to a five-year-old, or for that matter, to octogenarians.

Once germ cells are mature, they must be *released* at the right time, i.e. while both gametes are *mature* (fully developed), and in the correct number. For example, the release of sperm during intercourse will not lead to a fertilized egg if the woman is not ovulating, that is releasing the egg, at roughly the same time plus or minus a few days. Furthermore, even if she were ovulating during that time, the egg would not be fertilized if insufficient sperm were ejaculated to make the tortuous journey through the vagina, uterus, and oviduct so that some finally reach the egg. Those neural and hormonal factors affecting the formation and release of germ cells will be discussed in Chapters 3 and 5.

Transport and Union of Gametes

Finally, the gametes must be *transported* through the unobstructed male and female genital systems so that they meet under the most favorable circumstances. If a woman has "blocked tubes," for example, she may be considered sterile even though she can produce viable and mature eggs. Likewise, a man who cannot transport his sperm because of a vasectomy is sterile, although he still retains his "virility," his "manliness," that is his ability to carry on sexual intercourse.

But simple transport of mature gametes is not enough. The gametes must live long enough for fertilization to occur, that is they must be viable. Sperm usually remain viable in the female genital tract for up to two days, whereas once an egg is released, it remains

viable for 24 to 36 hours. Thus, for fertilization to occur, it is important to know when the egg is released and for how long both sperm and egg remain viable. On the other hand, in the "rhythm" method of contraception, it is important to know these same facts so that fertilization will not take place. In practice, it is very difficult to know the exact time that the egg is released.

Some Characteristics of Mammalian Reproduction

Human beings belong to the group of animals called *mammals*, which includes cats, dogs, lions, elephants, and apes; we share with them a number of common characteristics regarding sexual reproduction. For example: (a) All mammals fertilize their eggs *internally*, rather than externally as occurs in frogs and most fish. (b) The young develop inside the mother and make use of a transient organ, the *placenta* (Chapt. 22), that is situated between the mother and developing child. (c) The *gestation* period, that is the time between fertilization and birth, is prolonged, being one month for rabbits, nine months for humans, and a little over eleven months for horses. (d) The young are born *incompletely developed* by live birth; man's brain, in particular, does not develop fully until at least the age of five. (e) Finally, as the name mammal implies, members of this group possess *mammary glands* to provide their young with nourishment during their early development. Even male mammals have remnants of mammary glands, but they do not function to produce and deliver milk. The properties of the male reproductive system are discussed in the next chapter.

Chapter 2

Male Reproductive System

zygote = fertilized egg

Reproductive systems have two major functions: to *form gametes*, i.e. *sperm* or *eggs*, and to *deliver those gametes* so that they can unite to form the fertilized egg (*zygote*). In males, the sperm are formed by the testes, and they are delivered to the egg by means of the penis. Hence, in this chapter we will learn about the structure of the testes, the pathway that the sperm take to get to the penis, and the physiology of erection so that the sperm can be deposited within the female's reproductive system. In addition I will discuss: the role of certain glands in secreting a variety of materials which function mostly to keep the sperm alive; the structure of sperm; and fertility in males.

The Testes

The *testes* (the male gonads, also called testicles) serve as the primary sexual organs in the male genital system. They have two functions: one is to produce viable and motile sperm contained in a fluid medium. The other is to serve as the major site for producing steroid hormones (generally called *androgens*, e.g. testosterone) that affect the development of males and their secondary sex characteristics, such as facial hair and pitch of voice.

The testes are oval shaped organs, averaging 1 inch in diameter and 1 1/2 inches in length. They develop as a pair in the body cavity, and, just before birth, each descends through the *inguinal canal* into a thin-walled sac of skin, the *scrotum*. Once descended, the canal seals off. In cases where it does not seal off completely, or it tears, an *inguinal hernia* can develop. An undescended testis cannot produce viable sperm, but it can produce androgens.

There are cases of men having no testes. They may be removed for medical reasons, by accident, by acts of barbarism, or, in one case out of 50,000, an individual may be born without them. Can such individuals father a child? In at least one case it has been possible. Through the delicate process of microsurgery a testis from one identical twin having two testes was transplanted to the other twin, who lacked both testes. The recipient fathered a child.

Other than affecting the ability to produce sperm, the absence of testes affects adversely the development of the male secondary sex characteristics, such as facial hairs and a deep voice, which normally would be stimulated by the testosterone produced by the testes. Because women were not permitted to sing in church choirs during medieval times, the church sanctioned the castration of boys who had not reached puberty. These eunuchs, also called "castratos," retained their higher pitched voices and sang the soprano and alto parts in the choirs.

The testes of adult human males are in the scrotum because the testes require a temperature lower than that of the body in order to produce sperm. Such is not the case in all mammals. For example, the testes reside inside the body cavity of whales and elephants; bats, on the other hand, lower the testes just before the breeding season. In sexually mature humans there are special muscles in the scrotum which bring the testes closer to or further from the body, depending upon such factors as the temperature or the individual's emotional condition.

As regards higher temperatures inhibiting sperm production, some societies believe that hot baths, or the wearing of tight pants to keep the testes closer to the body, serve as temporary male contraceptives.

Components of Testes

The testes have a number of components worth understanding. Each testis is composed of about 100 small *compartments* [lobules] (Fig. 2-1). Within a compartment are about 1-2 feet of coiled *sperm-producing* [seminiferous] tubules (Fig. 2-1). These tiny *tubules* are made up mostly of two kinds of cells, the *pre-sperm* cells [spermatagonia] and the *nurse cells*, which nourish the developing sperm (Fig. 2-2). Once the sperm are almost completely formed, they are passed into the central cavity of the sperm-producing tubule where they will be transported further along the genital tract.

Situated in the interstices between the sperm-producing tubules, but still within the compartments, are the *interstitial* cells (i-cells) (Fig. 2-2). These are the cells that produce the androgen hormones. Thus, we see here an example of *division of labor* of three different cells within one organ: one forms sperm, another nourishes developing sperm, and a third produces hormones.

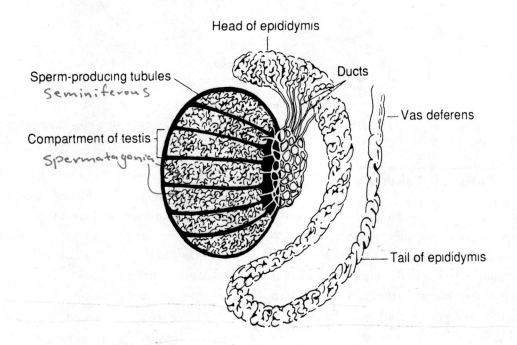

Head of epididymis

Sperm-producing tubules
Seminiferous

Ducts

Vas deferens

Compartment of testis
Spermatagonia

Tail of epididymis

Figure 2-1 Testis and associated structures.

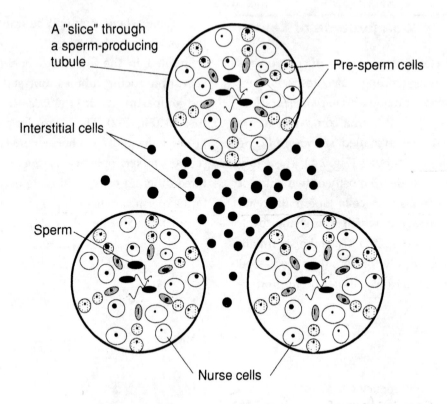

Figure 2-2. Cells of the testis. Each of the three large circles represents a slice taken through a sperm-producing tubule.

Pathway Taken by the Sperm

After the developing sperm leave the sperm-producing tubules of the compartments, they go through a network of larger tubules which all empty into a still larger coiled tube known as the *epididymis*, which serves as a storage and selection chamber (Fig. 2-1, 3). It takes about three to six weeks for each sperm to go through this coiled tube, which can reach a length of 15 to 20 feet were it to be stretched out into a straight hollow tube. It is in this chamber that unfit sperm are removed, and here that the remaining ones mature and are stored. The unfit ones, as well as the unused mature ones, are essentially dissolved and absorbed. The products of the dissolved sperm are used as a source of nutrition for adjacent tissues and cells. The epididymis also produces some fluids which eventually end up in the semen (p.15).

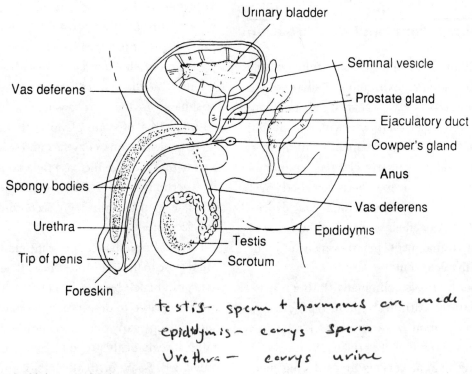

Urinary bladder

Vas deferens

Seminal vesicle

Prostate gland

Ejaculatory duct

Cowper's gland

Anus

Vas deferens

Epididymis

Spongy bodies

Urethra

Tip of penis

Foreskin

Testis

Scrotum

testis - sperm + hormones are made
epididymis - carrys sperm
Urethra - carrys urine

Figure 2-3. Male reproductive organs.

Vas Deferens

The mature sperm along with some of the epididymal fluids enter a long (ca. 18 inches) muscular tube known as the *vas deferens* (Fig. 2-1, 2). The word "vas" means a tube or duct, and from the cutting of this duct, we get the word *vasectomy*. This tube carries the mature sperm out of the scrotum into the body cavity (Fig. 2-3). The vas deferens is not the only link between the testes and the body cavity. The vas deferens, together with some nerves, arteries and veins, is the major link between the body cavity and the testes of the scrotum.

The last part of the vas deferens in the body cavity becomes somewhat muscular, and is called the *ejaculatory duct*. This duct passes the semen into the central tube of the penis known as the *urethra* (Fig. 2-3). The urinary bladder also passes urine into the urethra, but not during the ejaculation of semen.

Epididymis and Masturbation

Although the epididymis is fully capable of absorbing all extra unused sperm, there are a number of ways that stored sperm are removed other than by ejaculation during sexual intercourse. Frequently young men after reaching puberty ejaculate spontaneously during sleep. These so-called "wet dreams" usually stop with sexual experience, but nonetheless are often the cause of a certain amount of embarrassment and concern in adolescents.

Sperm are also eliminated through masturbation. Although such a practice is frowned upon by a number of religions, studies show that it is relatively universal among men at various times during their lives. Despite tales to that effect, there is no evidence that masturbation affects the body adversely.

Accessory Glands

Three glands, known as the seminal vesicle, the prostate gland, and Cowper's gland are all found in the body cavity, and contribute their secretions to the ejaculatory duct or the urethra (Fig. 2-3). It is still not quite certain what the secretions of each of these glands consist of, and what their specific functions are. The secretions of the seminal vesicle are thought to activate the swimming movements of the sperm's tail, whereas prostate secretions are believed to make the semen more alkaline (i.e. non-acid) so that the more acidic vaginal fluids will become hospitable to the acid-sensitive sperm. In addition, the prostate gland is thought to secrete calcium ions, enzymes, and some nutrients that serve to increase the viability of the sperm. Both the seminal vesicle and the prostate gland empty their secretions into the ejaculatory duct.

The chestnut-shaped prostate gland, located close to the large intestine and near the anus, can be felt by rectal examination; this method is used to detect cancer of the prostate. Through early detection, the life-threatening threats of this type of cancer can be prevented. Some of treatments of prostate cancer have a side effect that concerns most men, the loss of the ability to have erections.

The smallest of the male accessory glands, the pea-sized Cowper's gland, just before ejaculation secretes an alkaline fluid and "lubricant" into the urethra which allows for the safe passage of the semen. Occasionally, during sexual excitement stimulated by foreplay, the Cowper's gland may secrete a "love drop" which contains some sperm leaking over from the ejaculatory duct. Thus, although rare, there remains a possibility of pregnancy when the withdrawal method of contraception is used should the Cowper's gland release its secretion.

The Semen

The *semen*, or *ejaculate*, is composed of about 10 percent sperm and 90 percent of the combined secretions of the epididymis and the three accessory glands. Semen varies in consistency, sometimes being watery, at other times almost gelatinous. One ejaculate can contain from 200,000,000 to 700,000,000 sperm. This number is not as startling as it may seem, considering that the average male after he reaches puberty can produce approximately 500,000,000 sperm a day, or 50,000 per minute. In contrast, a female at most produces approximately 500 mature eggs in a lifetime.

The Penis

The copulatory organ of males, that is the organ which transmits the semen, is the penis (Fig. 2-3). In addition to transmitting semen, the penis also serves in the sexual gratification of the male, making sexual intercourse pleasurable, and thereby serving the species by encouraging sexual reproduction.

The size of the penis, which varies among men, has little to do with sexual gratification of either the male or female during sexual intercourse. Nevertheless, the penis of *Homo sapiens* is larger per body size than the penis of all other primates. Some anthropologists explain this relatively large size of the human penis as an evolutionary adaptation to attract female mates and thereby encourage copulation.

In a similar manner, the breasts of a female of the human species are enlarged all the time after she has reached puberty, whereas the females' breasts of other primate species enlarge only during lactation. The enlarged human female breasts, some anthropologists presume, also serve as a visual attractant to encourage the coming together of the two sexes so that they will reproduce the species.

The Process of Erection

The structure of the penis is designed to facilitate its hardening as occurs during an erection so that copulation can occur (Fig. 2-4). During an erection, i.e. the *tumescence* of the penis, blood enters the penis by means of *arteries*, and leaves the penis by means of *veins*. There are three arteries, each passing through one of three long *spongy rodlike erectile bodies* that traverse the length of the penis. Each of the spongy tissues is surrounded by a thin layer of leathery connective tissue. You should note that the lower

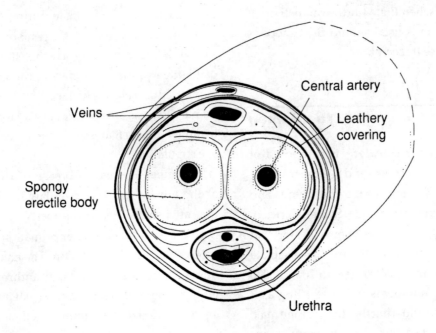

Central artery

Veins

Leathery covering

Spongy erectile body

Urethra

Figure 2-4. Components of penis involved in erection.

spongy body also contains the urethra, through which both urine and semen flow. The veins are located at the top of the elongated penis, just above the spongy erectile tissues.

If blood enters and leaves the penis at the same rate, the penis will not become erect, but will remain flaccid or limp. For an erection to occur, the inflow of arterial blood must be greater than the outflow of blood through the veins.

Here are the sequences of events that lead to an erection. First certain nerves stimulate the release of substances that widen (*dilate*)

the arteries of the penis. More blood flows into the penis and fills up the three spongy rodlike bodies. As the spongy bodies swell up, they squeeze against the veins of the penis, causing their diameter to become smaller. Thus, the outflow of blood through the constricted veins is reduced, the volume of blood in the three spongy bodies remains at an increased level, and the tumescence is maintained.

The erection ends as soon as the arteries return to their normal smaller diameter. When this happens, less blood goes into the spongy bodies, the pressure is removed

MORE BLOOD GOING INTO THE PENIS THEN OUT CAUSES AN ERECTION.

from the veins which then widen, and the blood flows out of the veins again at the same rate at which it enters.

Thus, the process of erection is a simple hydraulic one. It is not maintained by voluntary muscles, i.e. you cannot will an erection. There is no bone in the human penis, although a few mammals do have them.

The tip of the penis, called the glans, is an extension of the lower spongy erectile body, the one that contains the urethra. The glans has a circumference somewhat greater than that of the shaft of the penis. The glans is also rich in nerve endings, which when stimulated, are the source of sexual excitement and can lead to an erection, if there is none present already. The glans is usually covered with a piece of skin called the foreskin [prepuce] (Fig. 2-3). It is loose, like much of the skin surrounding the penis, so that it can accommodate the larger size of the erect penis.

Circumcision

There is some controversy about the desirability of surgically removing the foreskin (circumcision) as is routinely done among observant Jews, Muslims, Ethiopian Christians, and other peoples. Unless done correctly, there is some danger to the male, especially if done on adult males. On the other hand, with the foreskin removed, there is less chance of bacteria and other contami-

nants collecting in the *smegma*, a secretion of the *Tyson gland* which collects between the foreskin and tip of the penis. Recent reports indicate that circumcisions prevent urinary tract infections in infancy, cancer of the penis in adult men, and the transmission of a number of venereal diseases. In addition, it appears that sexual partners of circumcised males have lower rates of cervical cancer. Today circumcision is a relatively common procedure in American hospitals, although it was more common twenty years ago.

Impotence and Implant Surgery

For men who are impotent because of a poor circulatory system, there is a surgical procedure that reconstructs an artificial hydraulic mechanism which will allow them to have erections. Such a device is present in at least 5,000 men today. It is an inflatable surgically implanted hydraulic apparatus that insures an erection at any time desired. In America alone there are probably 5,000,000 physiologically impotent males who could profit from this device. It is not recommended for males who are psychologically impotent; they need psychiatric and counseling treatment, which is less expensive and less painful than implant surgery.

The device is quite simple and merely mimics the biological events of blood flow. Instead of blood flowing into the penis, however, there is a miniature pump in the scrotum near the testis that pumps in a special fluid. The pump is connected to a reservoir of fluid, which is stored in the body cavity, and which is connected to two inflatable cylinders that have been surgically implanted along the length of the penis. Before intercourse, the individual need merely turn on the pump that will inflate the cylinder. Men with these implants may have normal sexual relations and father children.

Some newer research indicates that the bloodflow to the penis can be increased by the injection of certain substances that cause arteries to dilate. Although this process is much simpler than having the implant operation, it does have the disadvantage of requiring an injection each time an erection is desired. Research on cures of impotency is continuing, however, because of the widespread incidence of this disorder.

Impotence — Psychological or Organic?

I have just said that the hydraulic device is not recommended for patients who are psychologically impotent. How do we determine whether or not a person is psychologically impotent? The answer lies in measuring the frequency of erections that occur during sleep (*nocturnal penile tumescence*, or *NPT*).

The NPT test for psychological impotence is based on observations that during an 8-hour sleep, the average healthy male has about three to five erections, sometimes lasting a total of 2 hours. By monitoring the NPTs of impotent patients, doctors get insight into the nature of the type of impotence. For example, if a patient has no NPTs, or if the duration of the NPTs is significantly reduced, then perhaps these individuals have some sort of hormonal imbalance, a problem in their blood flow, or some obstruction of the penile artery. Diabetics, for example, because they have blood vessel obstructions, have a high degree of organic impotence.

A male who is psychologically and not organically (physiologically) impotent, on the other hand, has NPTs of normal frequency and duration. Apparently the emotional factors that inhibit his ability to have erections during the waking hours do not operate while he is asleep. Hence, it is easy to see why NPT measurements are important in diagnosing impotence and blood flow problems, and in prescribing artificial hydraulic implant surgery.

Structure of Human Sperm

The mature human sperm has only about 1/70,000 the volume of the human egg. It is shaped like a miniature tadpole (Fig. 2-5) that has three major parts: the head, middle piece, and tail. The head contains the genes which the father contributes during fertilization, the tail propels the sperm to the egg, and the middle part provides the energy to the tail.

The entire sperm (Fig. 2-5), like all animal cells (Fig.2-6), is covered by a thin *cell membrane*. This membrane, which is made up of protein molecules floating in a layer of fatty material, regulates the flow of food sub-

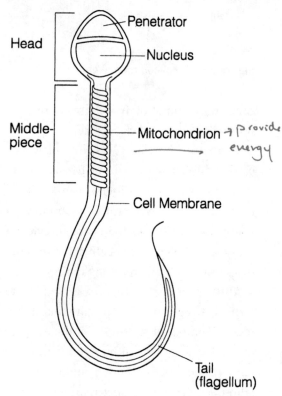

Figure 2-5. Basic morphology of a sperm.

stances, such as oxygen and sugars, and of wastes, such as carbon dioxide and urea, in and out of cells. The sperm is so small that it does not have much storage space for food; hence, it relies upon its nutrients coming from the fluids surrounding it and passing through the cell membrane.

The main component of the head is the nucleus, which, like the nucleus of the egg, contains exactly half the number of chromosomes found in the fertilized egg and in all other cells of the body. The chromosomes, which are thread-like bodies, contain the genes, i.e., the material containing the genetic information that the father transmits to the child.

At the tip of the head of the sperm is the *penetrator* [acrosome], a structure found only in sperm. When the penetrator touches the surface of an egg, it will initiate a series of reactions which will allow the sperm's nucleus to enter the egg's interior. Notice that I said allows the sperm's nucleus to enter. I did not say to allow the entire sperm to enter. More about that in Chapter 17.

The tail is made up of a long whip-like structure called a *flagellum*. You may recall that the word "flog" refers to whipping. Even the word "flag" refers to the whip-like motion of the flag in a wind. The flagellum has the same internal structure of the eyelash type of hairs (cilia) found along many of our cells, along gills of fishes, and at the tip of our sensory cells. It is the whip-like motion of the sperm's flagellum that propels it.

The structure of the middle of the sperm that makes energy for the movements of the tail is called the *mitochondrion*; it is the "powerhouse" of the cell. Mitochondria are found in all animal cells (Fig. 2-6). They burn sugars and other nutrients in order to make a soluble substance known as ATP. It is this ATP that is the direct source of energy for the tail to move.

In summary, the middle piece provides the energy for the tail to move the entire sperm so that the nucleus, by means of the penetrator, can enter the egg.

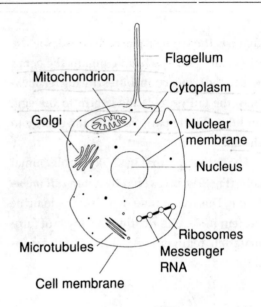

Figure 2-6 Simplified animal cell.

Sperm and Sterility

The adjectives infertile, impotent, sterile, and barren are applied to individuals who cannot father or bear children. We often use "barren" when referring to women and "impotency" in reference to men. Infertility and sterility usually refer to either males or females who cannot have children of their own, whereas the term "sterilize" refers to the process of rendering a person infertile.

Nonetheless, a person who is infertile may produce healthy gametes. A woman, for example, may produce normal eggs, but not be fertile because her tubes (oviducts) are blocked. Likewise, men may produce viable sperm, but be unable to father children be-

cause they cannot deliver those sperm by a normal erection and ejaculation.

There are cases, however, in which men can produce viable sperm and can have erections, but still be considered sterile. These cases usually derive from the amount of sperm delivered and the characteristics of those sperm. Both factors affect the number of sperm eventually reaching the egg.

Although only one sperm is needed to fertilize an egg, an ejaculate contains an immense number of sperm — on the average about 500,000,000 — although this number varies greatly per individual and per ejaculation. If many fewer sperm are ejaculated into the vagina, then chances of fertilization occurring are greatly reduced. A person producing 50,000,000 sperm or less per milliliter of semen (30 milliliter make up one fluid ounce) is considered infertile. Infertility in males can also occur if 25 percent of the ejaculated sperm are abnormal, or if the sperm clump together and do not swim freely.

Low sperm counts caused by exposure to hazardous chemicals in work are being recognized today as significant causes of male infertility.

Why are so many sperm required for fertilization? When you consider the very small size of the sperm, and the distance in the female's reproductive system between the site where the sperm are deposited and the place where fertilization occurs, the sperm take an incredible and difficult journey. First, the route is long and tortuous (Fig. 4-1), going from the vagina, through the uterus, and up almost two thirds the length of the oviduct. Secondly, the sperm must pass through narrow openings to get into the uterus and then into the oviduct. Thirdly, the proportion of sperm in the semen is diluted even further by the female's secretions. Finally, the acidity of the fluids in the vagina is hostile to the sperm. Nonetheless, if sufficient sperm are ejaculated, 100-200 may perhaps reach the unfertilized egg in the oviduct; only one will fertilize it.

It takes but a few minutes for the sperm to reach the uterus, and another hour before some get into the oviducts. The sperm which get that far can live for 2-3 days there. Because the egg must be fertilized within 1-2 days after entering the oviducts, there are only about 4-6 days at most between menstrual periods during which it is possible for the egg to be fertilized.

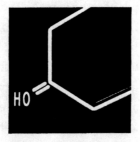

Chapter 3

Hormonal Control of the Male Reproductive System and Introduction to Contraception

The onset of puberty, the production of sperm and eggs, the development of the secondary sexual characteristics, such as enlargement of the breasts in females, and growth of facial hair in males, are but a few of the many events in our body regulated by *hormones*. Hormones, in fact, together with the nervous system, regulate most of the vital processes of the body. For this reason alone we ought to have some knowledge of hormones and their actions.

Yet modern science and technology have given us, for better or for worse, another reason making it imperative that we understand their actions: the commercial as well as illegal availability of a wide variety of hormones. Therefore, before you start taking hormones, you had better know exactly what they are and what they do.

Definition of Hormones

Hormones are substances that regulate a wide number of important processes in the body. Insulin, for example, regulates our sugar metabolism. Testosterone regulates a wide number of masculine characteristics, such as the presence of facial hair and the absence of cranial hair (baldness), the development of a deep voice, and the formation of a more robust bone and skeletal structure. We will be discussing mostly endocrine hormones, that is, hormones made in one part of the body which are transported via the circulatory system to exert their actions in other parts of the body. Those tissues and cells responding to hormones are called *target* tissues and cells. For example, an *endocrine* hormone known as FSH (see below) released from the pituitary gland at the base of the brain acts on target tissues in the testes to stimulate the initiation of sperm production.

An analogy of how endocrine hormones act on target tissues may be found in a consideration of how radios operate. Consider the *radio transmitter* as a gland that, instead of sending out hormones, sends out a number of radio signals. The signal, like the hormone, may be recognized by a number of target tissues, or, in the analogy, by a number of radio receivers set at the correct frequency. In our analogy, the physiological effect of the hormone, i. e. the response of the target, will be an amplified signal, whether it be a news broadcast, advertisement, or music.

That is not all. Just as sufficient hormone must be released into the bloodstream to reach the target tissue(s), the signal transmitted must be powerful enough to reach the receiver. Furthermore, just as the target cells or tissue must possess the proper receptor to recognize the hormone, the receiver must be set at the proper frequency to receive the signal. Finally, just as the target tissue must be functional in order for the hormone to be expressed, the radio must be on and the signal amplified.

General Characteristics of Hormones

Hormones regulate the rates of certain critical processes of cells and tissues. Only small amounts of hormones are needed to produce major effects in the body. Eventually the hormones are broken down in the body and the by-products are excreted in the urine. Hence, hormones must be produced at a rate coordinated with body needs. Some hormones may affect the production and release of other hormones. Many of these latter pivotal hormones are released from the pituitary gland.

Pituitary Gland

This endocrine gland is the most influential one in the body. It secretes a number of hormones that regulate the production and release of many more hormones. Among the body activities it controls are those related to producing both gametes and those steroid hormones which affect our secondary sexual characteristics.

The *pituitary*, the so-called "master gland," is a small pea-shaped gland attached by means of a stalk to the base of the hypothalamic region of the brain (Fig. 3-2). It is activated by both nervous and chemical signals from the brain. This activation can lead to the release of about eight different pituitary hormones into the bloodstream, hormones which eventually affect most of the body's activities. The *nervous* stimulation activates the posterior part of the gland to release two major hormones, vasopressin and oxytocin. Oxytocin, as we will see later, greatly influences the birth process and the release of milk.

Of the six known major hormones produced by the anterior part of the pituitary, three are very important in understanding human reproduction: (1) follicle stimulating hormone (FSH), found in males and females; (2) luteinizing hormone (LH) of females, which, in males, is called interstitial cell stimulating hormone (ICSH); and (3) prolactin of females, also called lactogenic hormone.

Hormones Affecting the Male Reproductive System

At puberty, the hypothalmus of the brain begins to produce a substance, called *gonadotrophin releasing factor* (GnRF), which it passes on to the anterior pituitary (Fig. 3-2). The GnRF causes the pituitary to release two hormones into the bloodstream. One is called the follicle stimulating hormone (FSH), which controls sperm production, and the other the interstitial cell stimulating hormone (ICSH), which controls the production of testosterone by the interstitial cells of the testes. The hypothalmus passes the releasing factor directly to its target tissue, i.e. the anterior pituitary, by means of small blood vessels connecting the two.

Because the actual interactions of the releasing factor (GnRF), the pituitary hormones FSH and ICSH, the steroid hormones, and the brain are so complex and still not fully understood, I have greatly simplified my explanations of them. In so doing I have highlighted general principles, and those actions and interactions that are relevant to understanding how our reproductive systems work.

Hormonal Action in Sperm Production

Unlike hormonal regulation of egg production in which only one egg matures monthly, the pituitary hormones affect the continuous production of millions of sperm daily in the adult male. To help understand how the various hormones affect sperm production, follow Figure 3-2 as you read these paragraphs. We have already learned that the brain, by means of GnRF, affects the release of FSH and ICSH from the anterior pituitary into the bloodstream. The target tissues for the released FSH in males are the *sperm-producing tubules* [seminiferous tubules] of the compartments of the testes. The FSH stimulates the increase in size of those tubules and of the testes, and stimulates the early phases of sperm formation. The ICSH, on the other hand, stimulates the interstitial cells (i-cells) which lie between the sperm-producing tubules to secrete the steroid hormone *testosterone* (Fig. 3-1). This hormone has many functions, the most important one for present purposes being to maintain the process of sperm formation once it is initiated by FSH.

Testosterone also stimulates the prostate gland and seminal vesicle to produce some

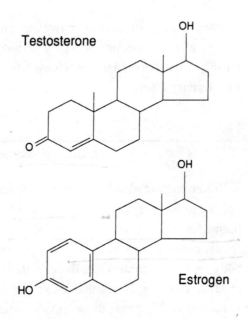

Figure 3-1 Chemical structure of testosterone and estrogen.

of the secretions making up semen. The hormone is also thought to affect the sex drive. During puberty testosterone affects the distribution of hair on the face and in other regions, the size of the penis, protein and muscle metabolism, the length of bones, as well as other *secondary sex characteristics* of men.

Interestingly, the i-cells of the testes also produce small amounts of a female hormone, *estrogen*. Similarly, the ovaries of females also produce small amounts of testosterone. From looking at the chemical structure of these two hormones, both of which have the so-called *steroid* shape (Fig. 3-1), you can see that they are almost identical.

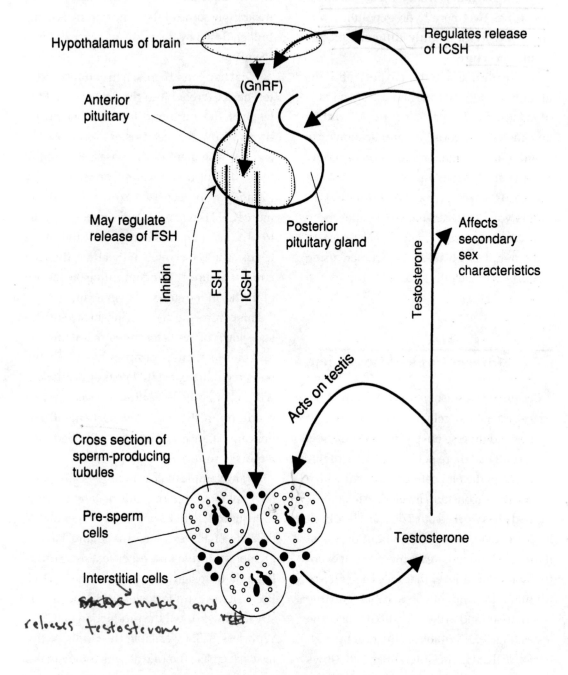

Hypothalamus of brain

Regulates release of ICSH

(GnRF)

Anterior pituitary

May regulate release of FSH

Posterior pituitary gland

Inhibin

FSH

ICSH

Testosterone

Affects secondary sex characteristics

Acts on testis

Cross section of sperm-producing tubules

Pre-sperm cells

Interstitial cells

makes and releases testosterone

Testosterone

Figure 3-2. Interactions of hormones of brain, pituitary gland, and testis.

Estrogen has 2 more hydrogen atoms than testosterone

The estrogen differs from the testosterone in that it has two more hydrogen atoms, and consequently, a slightly different though similar structure.

If the delicate balance between the amounts of testosterone and estrogen in the bloodstream of males is tampered with by any means, all sorts of complications can result, ranging from feminization of certain male characteristics to cancer. A word of advice to the wise: Do not take hormone pills unless you are under the strict supervision of a competent physician. Similar precautions ought to be taken by women using steroid hormone pills as contraceptives.

Interactions of Hormones

Because the testosterone released from the testes' interstitial cells has so many target tissues and diverse effects on the body, it is important that there not be too much of this hormone in the bloodstream. To maintain a *steady state* (constant) level of testosterone, the body has worked out a unique "feedback loop" method. When the amount of testosterone in the bloodstream increases, it eventually reaches a level that tells the anterior pituitary to send out less ICSH into the bloodstream and to the interstitial cells of the testes (Fig. 3-2). Without activation by high levels of ICSH, the interstitial cell slows

down its production of testosterone. In the meantime, some of the testosterone is being broken down by the body and excreted into the urine.

What happens, then, when enough testosterone is excreted into the urine so that the level of the hormone in the bloodstream drops below the steady-state level? As the level of testosterone decreases, the inhibitory effect of that hormone on the anterior pituitary also decreases, so that eventually more ICSH is released to stimulate the i-cells to produce more testosterone. If too much testosterone is secreted at that time, then the entire inhibitory cycle controlling the release of the hormone starts all over again.

This control mechanism in which a steady-state level of testosterone is maintained is just one of many examples by which the body regulates constant levels of substances within our body. We call such a steady-state condition in the body *homeostasis*. If that delicate balance is disturbed, the body attempts to restore it.

As an example of how homeostasis works, let's say we inject an adult male with extra testosterone. Will he become a "super male" because of the extra hormone and have a greater sex drive and other features stimulated by testosterone? Of course not. Why? Because the extra testosterone in the bloodstream slows down the production of ICSH. With less ICSH available the i-cells of the testes decrease the rate at which they make

homeostasis - Balance in body.

testosterone. As the excess testosterone is broken down and excreted, its homeostatic steady-state level is eventually reached and all is well.

On the other hand, let's say for medical reasons, that an adult male is given injections of a female steroid hormone. Will certain of his traits become feminized? The answer is "yes" for the following reasons. Just as does testosterone, some female steroids also prevent the pituitary from releasing large amounts of ICSH. Thus, in the male injected with female hormones, the female steroid inhibits the pituitary from releasing ICSH, the i-cells stop making testosterone, and, consequently, the *ratio* of female steroid to testosterone in the bloodstream increases. Such an imbalance of the amounts of male and female steroid hormones causes the male to take on a number of female characteristics, such as developing prominent breasts.

Anabolic steroids: These substances have made the headlines as they are not allowed to be taken by athletes competing in most national and international competitions, such as the Olympics. Anabolic steroids are synthetic hormones which retain primarily the anabolic, that is body building, properties of such natural steroids as testosterone. As a word of caution, anabolic steroids should be taken only on the advice of an experienced medical endocrinologist. These drugs may cause irreversible harm to the recipient depending upon the dose, the nature of the steroid, and the age and physiological condition of the recipient.

Inhibition of FSH production: In most cases, males produce sperm continuously throughout their adult life. Therefore, it would seem unlikely that sperm would produce another hormone to regulate FSH release just as the i-cell produces testosterone which regulates ICSH release. Nonetheless, there may be some subtle form of regulation taking place affecting FSH release. Some scientists believe that cells in the sperm-producing tubules produce a substance called *inhibin* that decreases the amount of FSH released. Future research may prove them right (Fig. 3-2).

Male Contraceptives

I have decided not to have a special single chapter on contraceptives as is customary in most books of this type. Instead, I have chosen first to describe the biological process or system on which the contraceptive acts, and then to discuss the particular contraceptive related to the topic under discussion.

Thus, now that we have discussed the male reproductive system, we are in a position to understand how male contraceptives work. Similarly, after we cover female hormones, we will discuss how "the pill" interferes with those hormones. And, after we discuss implantation of the early embryo into the wall of the uterus, we will discuss the intra-uterine device (IUD), which interferes with that implantation process. My rationale for this approach is my belief that in order to use contraceptives effectively and safely you need to understand how they work (and sometimes don't work) rather than simply follow some directions.

First, however, a few words about contraceptives in general. Contraception serves as a means to prevent fertilization while allowing sexual intercourse. Contraceptives may be used for personal reasons to restrict the number of children in a family or for societal reasons to control the size of populations. To achieve population control, various "civilizations" have used such other methods as infanticide, castration, abortion, barriers to early marriage, and a number of means to encourage abstention from intercourse, including financial rewards. The earliest records indicating the practice of some sort of contraception of which we are aware go back to 1850 B.C.

There are, however, only two things that we can say for sure about contraception: One, the only absolutely certain way to avoid having children is to abstain from sexual intercourse. Two, of all the animals, only the human species has devised ways to have sexual intercourse and not have children — that is, most of the time.

Interfering with Sperm Production — a Male "Pill"?

"The pill" (contraceptive taken orally) works on females to stop egg production because the hormones contained in the pill essentially mimic the normal way that the woman's own body hormones regulate the ovaries to produce only one egg per month. Males, however, have no such natural mechanism because they produce sperm continuously throughout their adult life. Hence, there appears to be no natural sperm-inhibiting mechanism for a "male pill" to mimic, with the possible exception of the relatively unknown hormone inhibin. Thus, there seemed little prospect for developing an oral contraceptive for males until recent years when Chinese scientists came up with *gossypol.*

Gossypol is a non-steroidal substance derived from the seeds of some cotton plants that appears to inhibit the production of sperm in male mammals including humans. The compound seems to act by inhibiting a special enzyme found primarily in sperm and testis cells, an enzyme necessary for these cells to make energy. Gossypol is inexpensive to produce, and its effects are reversible. For example, a study of 10,000 men showed that gossypol produced 99.89 percent sterility if the men took small doses of the drug daily for two months. After this initial period, the men needed to take only two doses per month to remain in this temporarily sterile condition.

There are still problems regarding gossypol. For example, high concentrations can cause circulatory problems, heart failure, malnutrition, diarrhea, and hair discoloration. Even at the lower concentrations, 13 percent of the men developed some weakness and 6 percent had atypical decreased appetites. Although much research needs to be done, gossypol, or a derivative of gossypol, promises to be an excellent candidate as a male oral contraceptive.

Despite the promise of gossypol, there is still some research attempting to develop a male steroidal oral contraceptives. Prospects at this time do not look good, however. Some scientists have tried using steroids that are similar in structure to testosterone (such compounds are called *analogs*). Two of these testosterone analogs do cut down sperm production, but they have one disadvantage — they also abolish the ability to have erections.

Research in the field of male oral contraceptives is in its infancy, and is still being explored.

Vasectomy

Today there exist two major ways for men to prevent their sperm from reaching the female reproductive tract during sexual intercourse. Both are mechanical mechanisms designed to interfere, not with the production of sperm, but with the transport of sperm.

Vasectomy is a surgical process whereby part of the vas deferens, the main tube connecting the testes to the penis, is removed. The operation is simple, relatively inexpensive, and usually irreversible. It is done in the following way: (1) Part of the scrotum is treated with a local anesthetic. (2) A piece of the vas deferens is drawn out of the scrotum and about one inch is cut out of it and discarded. (3) Both cut ends of the remaining vas deferens are sealed tight. (4) The separated tubes of the vas deferens are put back into the scrotum which is then stitched up. The entire operation is quite painless and takes about five minutes.

There are a few risks involved, however. Occasionally there are problems from infections and clots developing from vasectomies carried out poorly.

After the operation, it takes about two months until the man will no longer have sperm in his semen, because there are still sperm in the remaining vas deferens lying in the body cavity. These sperm eventually are ejaculated, or die and are absorbed by the body. Sperm are still produced by the testis after the vasectomy, but in smaller numbers. Since there is no way for them to escape from the epididymis, they eventually die, and are "digested" and reabsorbed there.

There appears to be no effect of vasectomy on the sex drive and on the secondary sexual characteristics, all affected by testosterone. This hormone continues to be produced by the i-cells of the testis which are unaffected by the operation.

The one aspect of vasectomies that is of greatest concern to those considering the operation is its irreversibility. After vasectomy, the individual is essentially sterile and will therefore never be able to father a child again. Even though most men who have vasectomies do so because they feel they have sufficient children, there always remains the possibility that some may wish more children in the future. For example, they may lose their present children through some tragedy, or they may remarry and wish to start a second family.

Through a very painstaking, difficult process called *microsurgery*, some doctors are able to reunite the two cut ends of the vas deferens. This process, however, is not guaranteed to work, it is expensive, and relatively few specialists can do it.

There is another possible alternative for men who wish vasectomies and yet feel that sometime in the future there may be a chance that they will want children of their own. That alternative is to have samples of their semen frozen and stored. The process of using frozen sperm for artificial insemination has been well worked out in the field of animal husbandry of domesticated animals, such as cows and horses. In fact, an industry has been built around this process.

Condom

The most widely used male contraceptive, the condom, is based upon the "barrier" method, i.e. a barrier impenetrable by sperm is placed around the penis to prevent semen from entering the female reproductive system during sexual intercourse. This method is a very old one, with early condoms being made of such materials as "fish skins" and silk and other fabrics. Rubber condoms were used as early as the eighteenth century by the court of Charles II. The major event which eventually led to the manufacture and wide use of rubber condoms occurred in 1844 when Goodyear invented the process for vulcanizing rubber.

Today the use of condoms is widespread among men in countries where they are available. When used properly — and in time — they are extremely effective contraceptive devices. There are no harmful side effects to the user, as there are with most other contraceptives. Most important, condoms are also the best means, other than abstinence, for preventing the spread of venereal diseases because most infectious agents cannot penetrate the rubber barrier. In this respect, condoms are usually thought of as a protection for uninfected men. Similarly, they can also protect uninfected women from being infected by men harboring venereal organisms. Currently condoms are being recommended because of their relative effectiveness in lowering the spread of the deadly AIDS virus. A recent innovation which also may help in the prevention of the spread of the AIDS virus (HIV) is the *female condom,* also made of rubber, which the female can place in her vagina.

Chapter 4

The Female Reproductive System

Human beings belong to the group of animals called mammals and share the common characteristics of the reproductive system of those animals. One such major feature of female mammals is their small ovary which produces relatively few eggs. Mammals also have evolved an efficient reproductive tract which allows for internal fertilization and the transport of sperm to the egg, transport of the fertilized egg to the uterus, and a prolonged gestation period guaranteeing excellent conditions for the development and birth of the young.

In contrast, frogs or fishes, which are not mammals, produce millions of eggs at a time and, accordingly have gigantic ovaries occupying much of their body cavities. Fertilization of their eggs usually takes place outside of the body in a non-protective environment.

In this chapter, we will touch on only some of the descriptive and structural highlights of how human eggs are produced and the anatomy of the female reproductive tract. In chapters that follow we will get a much deeper understanding of the formation of eggs and of the hormones which control their development and release.

The Ovary

Ovaries exist in pairs (Fig.4-1), each one being about the size of an almond. They have two functions. One is to store, mature, and liberate eggs. The other is to produce hormones that, among many functions, affect the reproductive tract by: preparing the uterus for pregnancy, maintaining the

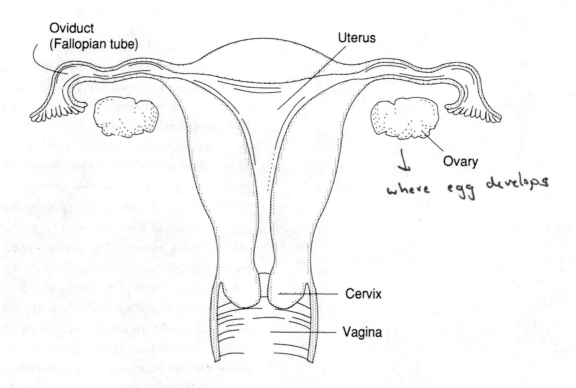

Figure 4-1 Female internal reproductive organs.

uterus during pregnancy, and signaling the anterior pituitary to secrete less FSH and LH.

Egg Formation

In the developing ovary of the fetus, there are millions of pre-egg cells, i. e. cells that have the capacity to become mature eggs. By birth, this number is reduced to about two million, and by late puberty, only about two hundred and fifty thousand pre-egg cells remain in each ovary. During an average lifetime from puberty through menopause, a female releases about four hundred to five hundred mature eggs. As will be seen in the chapters on fertilization, only a fraction of the pre-egg cells actually complete the entire process of *egg formation* (oogenesis).

As an immature egg (oocyte) develops in the ovary, it is surrounded by a cluster of cells. This entire package of oocyte and surrounding cells is called a *follicle* (Fig. 4-3). Starting at puberty, every month a number of follicles start enlarging. A cavity develops between the oocyte and most of the cells in the cluster, and becomes filled with fluid. At the end of almost two weeks only a few of these follicles become fully enlarged with fluid to become a *mature follicle* [Graafian follicle].

During this change from the immature to the mature follicle, the developing egg goes through the first of its divisions in which the number of its chromosomes is reduced from

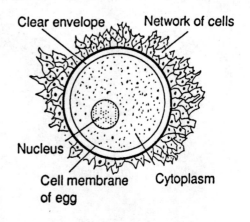

Figure 4-2. Human egg before fertilization.

forty-six to twenty-three. In addition, as the follicle matures, the cells of the follicle that surround the cavity start to secrete the female steroid hormones estrogen (estradiol) and progesterone (progestin).

Ovulation

Every month, one of the two ovaries releases an oocyte (egg) that was contained within one of the mature follicles. This process, called *ovulation*, alternates monthly between the ovaries. The timing of ovulation in the female's reproductive cycle is controlled by pituitary hormones. The egg which is released into the *oviduct* (uterine tube, Fallopian tube) (Fig. 4-2) is smaller than the period

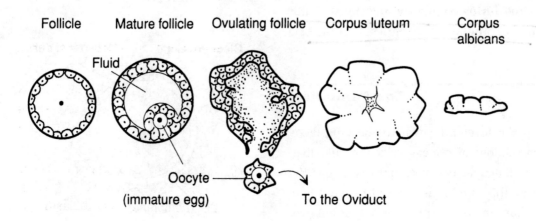

| Follicle | Mature follicle | Ovulating follicle | Corpus luteum | Corpus albicans |

Fluid

Oocyte

(immature egg)

To the Oviduct

Figure 4-3. Changes involved in egg production and ovulation in the ovary.

at the end of this sentence. Unlike the eggs of birds, snakes, and frogs, the human egg has little yolk to serve as a store of nutrients.

When the human egg is released, it is coated with a clear layer of a gelatin-like substance, called the *"clear envelope"* [Zona pellucida] which acts to hold the dividing cells of the fertilized egg together as it is received by the oviduct and travels down it. Attached to this clear layer is a loose *network of cells* [corona radiata] which sends out long extensions (Fig. 4-2). The extensions of those cells increase the surface area around the egg and may serve to enhance the take-up of nutrients from the surrounding fluids for the nutrient-poor egg. The released egg can survive for about twenty-four to thirty-six hours after it is released. If it is not fertilized within this period,

it dies and is absorbed within the female reproductive tract.

After the egg is released, a number of changes take place in the ovary during the ensuing two weeks. First, the ruptured follicle heals, some blood vessels grow into it, and it develops into a temporary endocrine gland, a "yellow body" called the *corpus luteum* (Fig. 4-3). The degree of temporariness of this gland depends upon whether or not pregnancy occurs. If it does, then the corpus luteum is retained during the childbearing months (*gestation period*). If pregnancy does not occur, the gland persists for thirteen to fourteen days, it degenerates, and menstruation follows. A whitish scar tissue, called *corpus albicans*, develops in this site of the degenerated corpus luteum. The nature and effects of the hormones released by the ovary are described in Chapter 5.

The Oviduct

On ovulation the egg is released into the oviduct (uterine tube, Fallopian tube) (Figs. 4-1,5), which conveys it to the uterus. An average oviduct is about four inches long and is lined with undulating little hairs (*cilia*) which help to move the egg along to the uterus. The tip of the oviduct (*fimbria*) lies close to the ovary. The fimbria has many finger-like projections which allow it to cup over the ovary when ovulation occurs to insure that the egg, as it is ejected from the ovary, gets into the tube of the oviduct. Just before ovulation takes place, the projections of the fimbria swell up with more blood and then virtually surround the ovary.

Usually the egg is fertilized while it is in the upper third of the oviduct. If the fertilized egg implants in the oviduct, or slips through openings between the ovary and the oviduct to implant in the female's body cavity, the potentially dangerous situation known as an *ectopic* pregnancy may develop. Normally, however, it takes about four to five days for the fertilized egg to reach the upper part of the uterus where implantation normally takes place.

The Uterus

The uterus, or "womb" (Figs. 4-1, 5), functions to receive the fertilized egg, to provide a site for it to implant, to nourish the developing embryo and fetus, and to expel the child during birth. It is a hollow, pear-shaped, muscular organ; in the nonpregnant female, it is almost two and one half inches wide and three inches long. It is located between the bladder and the end of the large intestine. The oviducts open into each of the two broad corners of the uterus. At the tip of the pear-shaped uterus is the *cervix* (Fig. 4-1), about one inch in diameter, which opens into the vagina (see Fig. 4-5).

The uterus has a thick muscular wall (Fig. 4-5) consisting of two layers (Fig. 4-4). The outer muscular layer (*myometrium*) is made up of involuntary muscles, that is, their contractions are not under voluntary control as are, for example, the voluntary muscles of the arm. The uterine muscles have two main functions. One is to allow for large stretching, especially that occurring as the fetus gets large in late pregnancy. The other permits the exertion of a large downward pressure needed to expel the child during labor.

The other layer of the uterus, the inner "menstrual" layer (*endometrium*) (Fig. 4-4). is the layer which thickens during the monthly cycle so that it can provide a nutritious environment for the implanted and growing embryo to grow and develop. It

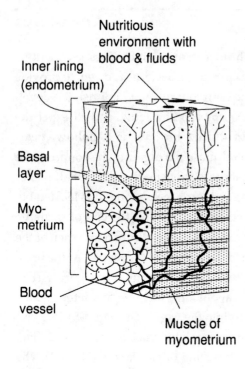

Inner lining (endometrium)

Nutritious environment with blood & fluids

Basal layer

Myo-metrium

Blood vessel

Muscle of myometrium

Figure 4-4 Piece of wall of uterus.

also is the layer that is mostly sloughed off once a month during menstruation.

The Cervix

The word cervix is derived from the Greek word for "neck." For example, cervical vertebrae are the bones in the neck. In the uterus, cervix refers to the "neck" of the uterus, i.e. the elongated part of the uterus opening into the *vagina*, where it extends about one half inch (Figs. 4-1,5). It is through this opening that sperm enter to get into the uterus, and through it the newborn exits from the uterus. The cervix is also the site in the reproductive tract from which cells are scraped in the "Pap" test for cervical cancer.

The cervix secretes a mucus. When the egg is released, the mucus becomes watery in consistency, making it easier for the sperm to enter the uterus. During pregnancy, the mucus forms a thickened plug which blocks the opening of the cervix, thereby helping to prevent the uterine environment from becoming infected with bacteria from the vagina. By examining the "stickiness" of one's cervical mucus, it is possible to get an indication of whether ovulation is imminent or has passed.

The Vagina

This structure, (Figs. 4-1,5,6), known also as the *birth canal,* is a muscular tube which envelops the tip of the cervix, and opens into the *vestibule* of the external genitalia (Fig. 4-6). This tube can distend greatly during delivery of the child. It is the structure which receives the penis during intercourse and which undergoes a number of contractions during orgasm. The vagina in humans, unlike in most mammals, is positioned at a right angle to the horizontal uterus. This right angle arrangement most likely evolved as a result of the upright walking position of

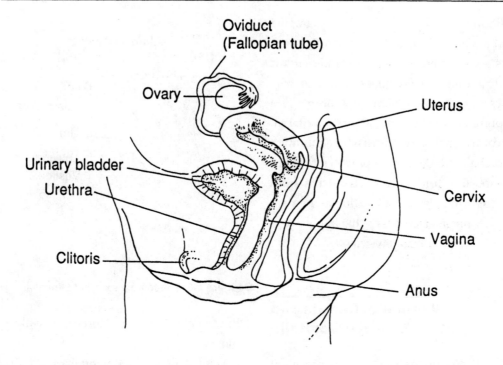

Figure 4-5. Female pelvic region, showing organs of reproduction.

human beings. It also may account for the so-called face-to-face position in sexual intercourse, a position rare among most animals except for humans and chimpanzees.

Although the vagina has no glands of its own, it does produce a secretion which serves as a lubricant during intercourse. Because the vagina opens into the external environment, it is subject to exposure to bacteria. The bacteria which grow in the fluids present in the vagina produce a weak acid harmful to sperm. The fluid in the male's semen is more alkaline ("anti-acid"), however, and can counteract the effects of the acids, thereby allowing more sperm to survive in the hostile acidic environment of the vagina.

The Hymen

This structure, called the *maidenhead* in days of past, has been the subject of numerous myths and taboos. It is a fold of a thin tissue that partially closes the external opening of the vagina. It can be ruptured by numerous kinds of physical traumas, such as by a hard fall, a shock to the pelvic area, sexual intercourse, or by a physician.

Because the absence of the hymen implies that a woman "has lost her virginity," various societies have placed a certain stigma and have inflicted cruelties on a woman who does not appear to show indications of a ruptured hymen during her first marital intercourse. In those cultures where this myth is prevalent today, some physicians, to aid prospective brides, can reconstruct from neighboring tissues an artificial hymen. Some women elect to have the hymen ruptured surgically by a physician.

Vestibule, Clitoris and other External Genitalia

The *vagina* opens into the *vestibule* of the external genitalia (Fig. 4-6). The vestibule is essentially a cavity bordered by the small inner lips (*labia minora*). At the upper end of the vestibule sits the *clitoris*, which is a small organ consisting of a stalk and a rounded tip. The clitoris develops in the embryo from the same tissues that become the tip end of the penis in males. The clitoris also has spongy tissues which become erect when the clitoris is stimulated. Like the tip of the penis, the clitoris is rich in sensory nerve endings. The clitoris is also covered by a skin, as is the penis. Some African cultures, even today, practice a custom on young girls similar to circumcision. In clitorectomy,

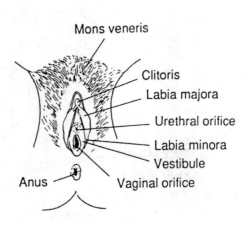

Figure 4-6 The female external genitalia.

however, the entire clitoris is usually removed.

The female external genitalia (called collectively the *vulva*), also contain the outer larger lips (*labia majora*) (Fig. 4-6). These are two prominent folds of a fatty tissue which have hair on the outside, and oil and sweat secreting glands on the inner surface. The outer lips develop in the embryo from the same tissues which form the scrotum in males (Chapt. 21).

Finally there is the *mons veneris* which contains pads of fatty tissue below the skin. The mons covers the region of the pubic bone, and it in turn is covered by hair. It is not difficult to see why the pubic bone and area are so named because the word *pubis* refers to hair. The mons contains a number of nerve endings which are stimulated by weight and during sexual excitement.

Path of the Sperm

From this brief look at the male and female reproductive tracts, we can see that the tiny sperm have quite a task to perform if one of them is to fertilize an egg in the upper part of the oviduct.

In the male the sperm are propelled from the epididymis in the scrotum through the long and narrow corridor of the vas deferens. Then they are catapulted by means of the ejaculatory duct through the penis's urethra into the inhospitable environment of the acidic vaginal fluids. The surviving sperm need to find the narrow opening of the cervix, after which they wander randomly in the cavity of the uterus. Some reach the tiny openings of the two oviducts. Those that enter the oviduct which does not contain an egg are, in practical terms,"lost" because they will never be able to have a chance to fertilize an egg. Those in the other oviduct, which for sake of discussion let us say contains a newly released egg, must swim upstream against a multitude of hairs (cilia) beating in the opposite direction. Finally, one sperm reaches the egg and fertilizes it. If another sperm reaches the egg, even a few seconds later, it cannot fertilize it. The door is essentially closed shut. Thus, it is no wonder that millions of sperm are needed for the process of internal fertilization.

IMPORTANT OVERVIEW (handwritten annotation)

43

Chapter 5

Menstrual Cycle and Female Hormones

There is perhaps no topic in biology which so many people think they understand and which so few do as that of the *menstrual cycle.* Once talked about in whispers in polite circles and usually never discussed in mixed audiences of men and women, this subject is one of the most complex and interesting in human biology. It entails the complex interaction of the brain, endocrine glands, ovary, and uterus which work together to produce one egg a month and a suitable environment to nurture the egg should it be fertilized.

To discuss the menstrual cycle fully and accurately would take one or more books. Given but two chapters to make the workings of the menstrual cycle understandable to the majority of us who are not trained research endocrinologists, I have taken two tracks. First, the actions of the hormones involved and their interactions with the brain are explained very simply; I do not dwell on the intricacies of hormone function which still are not fully understood and which continue to be the subject of current research. Second, I offer an overview of the subject together with the basic information and terms. Once I've provided you with the basic terms and processes, I then go over the same subject matter again somewhat more deeply, relating the scientific facts more directly to common experiences of most women.

Why Study the Menstrual Cycle?

There are a number of reasons: to satisfy scientific curiosity, to obtain information on personal hygiene, to understand the reproductive processes of the female body, to become psychologically prepared for womanhood, and to nurture understanding in males — whether they be husband or employer — of the differences in physiology of the two sexes.

To scientists, the study of the menstrual cycle is interesting for at least three reasons. The obvious one, which is the main subject of this book, is that it encompasses the biological interactions leading to the release of one egg monthly and the preparation of the

uterus for gestation of the fertilized egg until it eventually becomes a living human being.

Scientists are also intrigued with the study of the menstrual cycle because it provides us with a beautiful example of a most important property of organisms — that of biological cycles. For example, we now know that our hearts beat cyclically, i.e. rhythmically. Some of our glands secrete their hormones at twenty-four-hour intervals.

As another example, we sleep cyclically, i.e. about eight hours every twenty four-hour period. Those of you who have flown cross country or across the ocean have experienced jet lag. What is jet lag? It is the body's way of saying that although your watch reads 9:00 P. M., the body's internal biological clock is set so that it feels like midnight and the body must now rest. After a few days or so, depending upon the distance traveled, however, the body resets its internal clock so that the sleeping cycle coincides with that of the new place.

Thus, the menstrual cycle is an example of a cyclic process controlled by an internal biological clock in which the body releases an egg and readies the uterus for implantation of the fertilized egg. The average time for these cyclical processes is twenty-eight days. This cyclical nature of menstruation, i.e. the monthly "bleeding," was recognized early in history, and many myths and customs revolve around it. The word menstruation itself is derived from the Latin word *mensis* meaning month.

Finally, to the scientist, the menstrual cycle provides a striking example of how the brain exerts a strong influence on non-neural processes. Up until recent years, it was not clear how the brain controlled such processes as hormonal ones, and the study of the menstrual cycle provides us with insights into such control.

The other than purely scientific reasons for studying the menstrual cycle will be elaborated upon in the next chapter.

Phases of the Menstrual Cycle

We already know of the overall processes controlled by the menstrual cycle, i. e. egg formation and release, and preparation of the lining of the uterus for the fertilized egg. In general terms, menstruation is the cyclical process during which most of the inner lining of the uterus (endometrium) breaks down, and, along with some blood, is released into the vagina. The inner lining breaks down only if a fertilized egg does not implant in it. As one wag put it, the bleeding represents the "weeping of a frustrated uterus."

There are a few misconceptions that should be corrected now. For one, very little new blood from the circulatory system is released during menstruation. Most of the material comes from the shed inner lining; thus, the menstruation process does not represent a hemorrhaging. Secondly, although we speak of menstruation as a cyclical process of twenty-eight to thirty days, it can vary from twenty-one to ninety days depending upon the age, health, and physiological state of the individual.

There are four well-coordinated phases of the menstrual cycle: (a) The *destructive* (or *menstrual*) phase, defined as beginning on the first day of bleeding, refers to the breakdown of the inner lining of the uterus. It lasts an average of four to five days, but may vary. (b) During the next phase, called *proliferative* (or *follicular*) phase, the follicle(s) prepare for ovulation and the inner lining of the uterus starts to *grow* rapidly. This phase takes an average of ten days but can vary from six to sixteen days.

The key event of the menstrual cycle occurs during (c) the *ovulatory* phase, which lasts one to two hours. It is then that an egg is released from a mature swollen follicle. At about the same time, the mucous plug at the cervix starts to become more watery, thereby allowing sperm to enter the uterus more easily. Some women actually feel a slight short-lived pain during ovulation caused by a drop or two of blood from the ovary contacting the sensitive wall of the body cavity.

The fourth phase, called *secretory* (or *luteal*) phase, takes place very nearly twelve

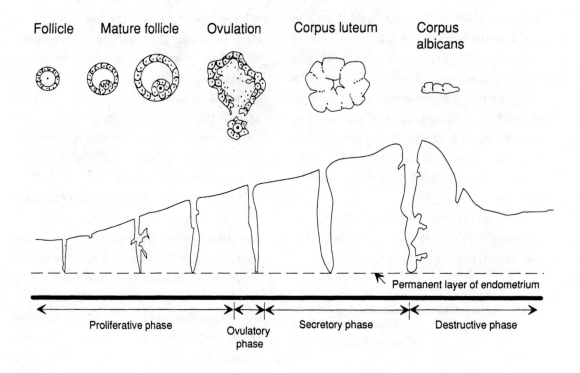

Follicle   Mature follicle   Ovulation   Corpus luteum   Corpus albicans

Permanent layer of endometrium

Proliferative phase   Ovulatory phase   Secretory phase   Destructive phase

Figure 5 - 1 Relationship between ovulation and the endometrium during the menstrual cycle.

to fourteen days after ovulation. In this last phase, the inner lining of the uterus thickens and forms small nutritive pools for the on-coming egg, the corpus luteum of the ovary secretes a large amount of female hormones, and the breasts swell. During this phase a woman may suddenly gain as much as five pounds of extra weight, which is just as quickly lost during the destructive phase. It is during the early part of the secretory phase that the egg can be fertilized.

The egg, either fertilized or not, is carried down the oviduct during the secretory phase and reaches the uterus. If the egg is fertilized, it will implant in the uterus and the menstrual cycle will not start again until some time after the child is born. If the egg is not fertilized, it will not implant and the destructive phase of shedding the inner uterine lining begins, and the entire menstrual cycle starts all over again.

Hormonal Control of Menstrual Cycle

Hormones Involved

There are three general categories of hormones involved in the menstrual cycle: (a) releasing factor(s), (b) gonadotrophins, and (c) steroids. The releasing factors are sent by the brain directly to the anterior pituitary. Thus far scientists are positive about the identity of at least one releasing factor (RF) affecting ovulation. Because this releasing factor stimulates the release of gonadotrophins, it is called gonadotrophin releasing factor, or GnRF.

The second category of hormones are called gonadotrophins (Gn) because they act directly on the gonads, i.e., in females acting on the ovaries and in males on the testes. The anterior pituitary releases three known gonadotrophins. One is called follicle stimulating hormone (FSH). Another is called luteinizing hormone (LH) in females and ICSH in males. Scientists are certain that GnRF from the brain stimulates the release of LH from the pituitary. It is less clear how GnRF affects the release of FSH. Until scientists discover the intricacies of the release of gonadotrophins by releasing factors, for our purposes we can state that GnRF controls the release of LH, and affects the release of FSH by some unknown complex means.

A third gonadotrophin released by the pituitary is called *prolactin* (also *lactogenic hormone*). This hormone seems to have many different functions. As the name implies, it is involved in milk production. It also inhibits egg production in humans. We will discuss prolactin in depth when elaborating on the failure to menstruate.

Not all gonadotrophins are produced by the anterior pituitary. For example, the placenta produces *human chorionic gonadotrophin* (hCG), a hormone which signals the ovary that the embryo has successfully implanted in the uterus. (The word chorion refers to the placenta which surrounds the embryo.) All gonadotrophins, in contrast to the compact cyclic steroid hormones (see Fig. 3-1), have a long coiled chain-like structure made up of amino acids (see Chapters 9 and 16).

The last category of hormones are the *steroids*. The ovary produces two main steroid hormones. One is *estradiol*, an *estrogen*, a word derived from the Greek meaning "to produce mad desire." This hormone is first produced by the developing follicle once it has been stimulated by FSH, and next by the corpus luteum. It is also produced by the placenta. For our purposes, we will use the terms estradiol and estrogen synonymously.

The other hormone produced by the ovary is *progesterone* (progestin), a word also derived from the Greek *pro gestare* meaning "to help with the bearing of young." The major source of progesterone is the corpus luteum stimulated by LH. Progesterone, like estrogen, is also produced by the placenta.

Sequence of Hormonal Control of Menstrual Cycle

These hormones have three main functions: a.) the releasing factors are needed to stimulate the release of the gonadotrophins; b.) the gonadotrophins effect the release of one egg monthly from the ovary; and c.) the steroids prepare the uterus so it will be receptive to and will nourish a fertilized egg. By keeping these points in mind, and by following Figure 5-2 as you read the ensuing paragraphs, you should have no problem in understanding the hormonal control of the menstrual cycle, and, for that matter, the means by which oral contraceptives (the "pill") act.

Starting at puberty, signals from the brain stimulate a major increase in the secretion of gonadotrophin releasing factor (GnRF), which then passes through small blood vessels directly into the anterior pituitary. There the GnRF stimulates the release of the gonadotrophins FSH and LH into the bloodstream. The target organ of those two hormones is the ovary. The FSH acts first by stimulating a number of follicles in the ovary

to grow and develop. Once the follicles start growing they produce and secrete estrogen into the bloodstream. The estrogen has two functions. First and foremost, it triggers the thickening of the inner lining of the uterus (endometrium), i.e. it stimulates the proliferative phase of the menstrual cycle.

The other function of estrogen is to regulate its own secretion by a sort of feed-back loop (Fig. 5-2) back to the brain and anterior pituitary. As the estrogen concentration in the blood reaches a certain level, it acts on centers in the brain and anterior pituitary to slow down or inhibit the release of more FSH. As less FSH is secreted, less estrogen is released by the follicles and a balanced (homeostatic) level of the two hormones in the bloodstream is reached.

About midway during the menstrual cycle, there is a surge in the release of LH from the anterior pituitary. By this time the follicles are already partially developed. The LH stimulates at least one follicle of the two ovaries to become fully mature and ovulate, releasing an egg into one of the oviducts. The LH then causes the ruptured follicle in the ovary to develop into a small yellow gland, the *corpus luteum.*

This gland, which persists as long as it is stimulated by LH, starts to secrete large amounts of progesterone as well as lesser amounts of estrogen. These hormones act on the inner lining of the uterus, with progesterone stimulating the now thickened lining

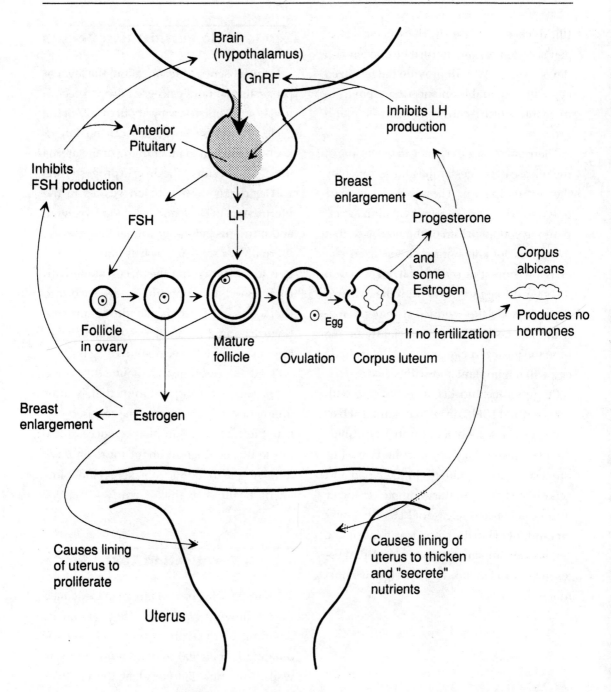

Figure 5 - 2 Female hormones during ovulation and the menstrual cycle. (Start by following dark arrow leaving "Brain.")

(the thickening caused by the action of estrogen secreted earlier) to further thicken and develop so that it will provide the fertilized egg with a favorable and nutritious environment into which it can implant, develop, and grow.

There is a sense of urgency in the timing of the release of progesterone and estrogen by the corpus luteum. These hormones have only about seven days after ovulation to prepare the endometrium of the uterus so that it may accept and nourish a fertilized egg arriving from the oviduct. It takes seven days for the egg to travel from the ovary to the uterus via the oviduct. If the endometrium has not thickened sufficiently by the time the fertilized egg arrives, the fertilized egg will not implant and will be lost.

Progesterone and LH, as is the case with estrogen and FSH, can be maintained at balanced levels in the blood. By about the middle of this *secretory phase*, a high level of progesterone accumulates in the blood and acts on the brain-pituitary complex to lower its production and release of LH. With lower amounts of LH in the blood to activate it, the corpus luteum starts secreting less progesterone until a balanced level of the two hormones is reached.

If No Fertilization Occurs

In the absence of fertilization, the level of progesterone (and estrogen) drops precipitously and gets so low by about day 27 of the cycle, that these hormones can no longer keep the thickened inner lining of the uterus intact. As a result of the drop in progesterone and estrogen, certain blood vessels in the uterus start to close and less blood, oxygen, and nutrients get into the inner lining of the uterus. Very soon afterwards, the lining begins to fall apart, and the pieces, along with some blood and other fluids, are shed into the vagina. When the shedding process (menstruation) is completed, the female begins the complete cycle all over again. She will do this monthly for about thirty-five years, except during and immediately after pregnancy. In summary, if the released egg is not fertilized and implanted successfully in the uterus, then at about fourteen days following ovulation the thickened inner lining of the uterus is shed.

If Fertilization Takes Place

Let us trace the path of an egg in the female tract following ovulation. The egg moves down the oviduct, propelled by the tiny undulating hairs (cilia) that are attached to its walls. This entire journey along the length of the oviduct to the opening into the uterus

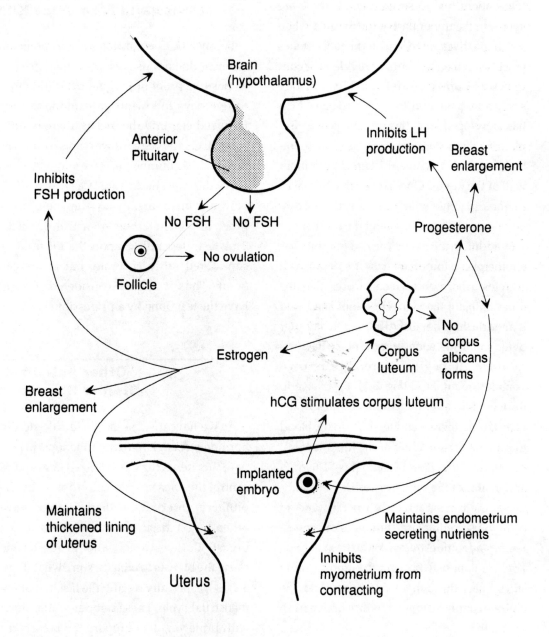

Figure 5 - 3 Actions of female hormones after pregnancy is established. (Start by following arrow leaving "Implanted embryo.")

takes about five to seven days. If there are sperm in the upper part of the oviduct when the egg arrives there, and fertilization takes place, then the egg starts to divide as it continues its journey down the oviduct. By day seven after ovulation, the fertilized egg which has developed into the pre-embryonic implanting stage (blastocyst), is capable of embedding itself into the thickened ripe inner wall of the uterus. Once it does that, it makes a simple placenta which serves as the connection between the embryo and the mother.

In addition, the newly-formed placenta has another very important task to perform. It secretes a hormone which insures that the inner lining of the uterus does not shed, and it must do this prior to the fourteenth day after ovulation. Thus, soon after the placenta starts to form, it begins secreting its own gonadotrophin, hCG (Fig. 5-3). By twelve to thirteen days after ovulation, and just in the nick of time, there is enough hCG in the blood to prevent the shedding of the inner layer of the uterus. How does hCG do this? The hCG acts in almost the same way as LH and stimulates the corpus luteum to persist and to continue secreting large amounts of progesterone and some estrogen. With these steroids now continuously present, the uterine lining stays intact throughout pregnancy and the embryo continues to get its nourishment from the mother.

Hormones and Pregnancy Kits

Because hCG is formed only in pregnant women, doctors use tests for this hormone as positive proof of pregnancy. Within four to five days following implantation of the fertilized egg into the uterus, there is sufficient hCG in a pregnant woman's urine to be detected. A number of years ago doctors used the so-called "rabbit-test" for hCG. Now a woman can buy an over-the-counter pregnancy kit to test her own urine for hCG. This test, depending upon the kit used, is considered fairly accurate. but errors can occur. Thus it always considered wise to have the test done by a professional.

Other Actions of Steroid Hormones

As we have discussed, the female steroid hormones serve two major functions: to prepare the uterus for the fertilized egg, and to control the release of gonadotrophins by the pituitary. In addition, these steroids have other target tissues. One of them is the breast. Both estrogen and progesterone cause the breasts to enlarge somewhat (Figs. 5-2,3,4), especially during the last half of the menstrual cycle. These same steroids, along with prolactin, also prepare the breasts for lactation. Since the prolactin levels needed for milk production are not secreted until near the end of pregnancy, however, milk itself is not produced during the menstrual cycle.

Progesterone has another important role to play in the uterus that becomes important should fertilization occur. It inhibits the spontaneous contractions that usually take place in the muscular outer wall (myometrium) of the uterus. If progesterone were not secreted during the gestation period by the corpus luteum and later by the placenta, uterine contractions might persist during pregnancy and cause the embryo to abort.

Other Hormones

We still are not certain of the role, if any, that prolactin plays in the monthly cycle. We do know, however, that *prolactin* inhibits ovulation (see p. 66), and if secreted in large amounts can cause the cycle to stop, leading to a condition known as *amenorrhea.*

Testosterone, a male hormone, may seem strange to mention while describing the female menstrual cycle. Not really. Recall that testosterone differs in structure from estrogen by a few hydrogen atoms and one oxygen atom off to the side (Fig. 3-1). In fact, these hormones are made both in the testes and in the ovaries by the same biochemical processes, except for the last few stages of their synthesis. It is not surprising, therefore,

to learn that the testes can make and release some estrogen and the ovaries some testosterone. The exact role(s) that testosterone, usually liberated by the follicle, plays in females is not thoroughly understood, although some believe testosterone affects the brain to increase the "sex drive" ("libido").

Steroids as Oral Contraceptives: "The Pill"

By now it may be apparent to you that if steroids can inhibit the release of FSH and LH, then it should be possible for steroids to serve as contraceptives. The search for contraceptives taken orally is not new. History shows the idea to be at least two thousand years old. Various cultures have concocted all sorts of herbal and other potions to prevent pregnancy.

Why did it take so long to develop effective steroid oral contraceptives when scientists knew of the actions of steroids in inhibiting ovulation years ago? There were at least two reasons. One was that steroids were expensive to make. The second was that steroids would not be absorbed easily into the body if taken orally. A breakthrough occurred which solved both problems when scientists discovered that a strain of the Mexican sweet potato made large amounts of a substance which was similar in structure to a steroid. This substance could be extracted cheaply from the sweet potato and be converted into forms of steroid hormones that could be taken orally, and that would act inside of the body just like estrogen and progesterone. Synthetic compounds that seem and act like the real ones are called chemical *analogs*.

Scientists then had to determine the correct sequence and the effective amounts of the two analog hormones to give in order to (1) mimic the actions of natural estrogen and progesterone in inhibiting ovulation, and (2) reduce other effects of the hormones. Today we have approximately four kinds of steroidal pills: (a) the *combination* pill; (b) the *sequential* pill; (c) the *minipill*; and (d) the *morning-after* pill. I will describe only the combination pill at this time because it fits in best with our discussion of the menstrual cycle hormones, and it is the most widely used of the female oral contraceptives today. For discussion of the three other types of contraceptive pills, see page 58.

Combination Pill

The combination pill contains a fixed proportion of estrogen and progesterone analogs to mimic approximately the levels of the natural hormones acting in the body. The estrogen analog inhibits the release of FSH,

and the progesterone analog inhibits the release of LH, just as high levels of estrogen and progesterone do naturally. Consequently neither the growth of the follicle stimulated by FSH nor the stimulation of ovulation by LH occur. Accordingly, without ovulation, fertilization cannot take place, and the pill is an effective contraceptive (Fig. 5-4).

Those of you who have been or are taking the pill, however, may not find this explanation satisfactory because you know that women on the pill do menstruate monthly. How could it be if FSH and LH, the initial triggers of the menstrual cycle, are not released? The answer becomes clear when you consider the following two points.

First, the pill itself consists of analogs of estrogen and progesterone, which are normally released from the follicle and corpus luteum that have been stimulated by FSH and LH. Thus, even though in women taking the combination pill no follicles mature, no ovulation occurs, and no corpus luteum forms, their bodies get the steroid hormones normally produced by the follicles and corpus luteum by swallowing them daily in the form of the pill. Accordingly the estrogen and progesterone analogs in the combination pill cause the uterus of the woman on the pill to form the same thick nutritive inner lining of the uterus that she normally would form during a regular menstrual cycle.

If you really understand what you are now reading, you should exclaim, "Aha! Now

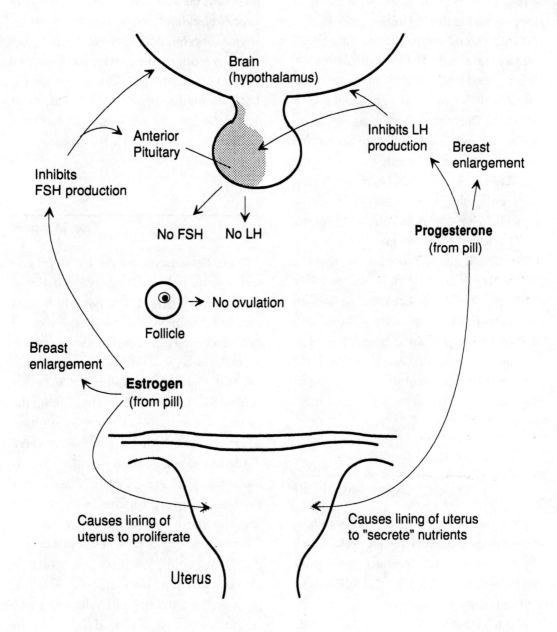

Figure 5 - 4 Actions of steroid (estrogen and progesterone) oral contraceptives. (Start by following arrows leaving "**Estrogen**" and "**Progesterone**.")

I've got you. How do you explain that if we keep on taking these estrogen-progesterone pills we should never menstruate, and yet when we are on the pill we do menstruate?" If you asked that question, then you are really with it, because it is a good and logical question. The answer is quite simple. In actuality, women on the pill take the steroid pills for only about twenty one or twenty two days during the cycle. On the remaining days, either they take no pills at all, or the pills taken are fake ones containing no steroids. Such fake pills are given only to simplify the pill-giving procedure so that the individuals do not have to calculate monthly when they should or should not go back on the steroid pills. Thus, during those days that no steroid pills are ingested, the estrogen and progesterone levels in the body drop considerably, and the inner lining of the uterus is shed just as in a normal menstrual cycle.

Sequential Pill

As its name implies, the dosages of hormones in the sequential pill are set to mimic the natural sequence in which the steroid hormones are secreted by the follicle and corpus luteum. For example, during the fifth through the sixteenth day, the daily pills contain mostly estrogen. Beginning on the seventeenth day and through the twenty-sixth day, the sequential pill consists of both progesterone and estrogen, with the progesterone concentration being relatively higher than the estrogen. In order for the menstrual flow to begin, the pills contain no steroid hormone for the remaining days of the cycle. To be effective, the sequential pill has to be in balance with the normal hormonal patterns of the user.

The Minipill

This pill consists of low dosages of progesterone only, and is taken daily, even during the menstrual period. The minipill is used mostly for women who react negatively to pills containing estrogen. It does not seem to act in the same way as does the combination pill, that is by inhibiting ovulation. Scientists believe that the minipill acts by making the cervical mucus hostile to sperm, by affecting adversely the little hairs (cilia) which carry the fertilized egg down the oviduct, by inhibiting implantation of the fertilized egg in the inner lining of the uterus, or by a combination of one or more of these effects.

Recently scientists have developed an implant containing progesterone which is placed under the skin surgically. This implant acts like the minipill, releasing small amounts of progesterone daily. Once in place, such implants can be effective for periods up to five years.

The Morning-After Pill

This pill is usually taken as the occasion demands; for example, it may be prescribed by physicians to women within a few days after they have been raped. It consists of a powerful synthetic estrogen known as DES (diethylstibestrol). If the dosage and the time it is taken are correct, DES can be very effective. It probably acts by affecting the inner lining of the uterus so that the fertilized egg does not implant and develop.

Other "Beneficial" Effects of the Pill

By taking the pill, women control the level of steroid hormones in their bodies as well as the time of the menstrual period. A survey shows that, for the most part, women on the pill as compared to women not on the pill do not have as severe menstrual cramps, have greater regularity in their menstrual periods, and lose relatively less blood and fluids during menstruation.

Disadvantages of the Pill

Steroids, recall, have other target tissues besides the uterus and the brain-pituitary complex. For example, during puberty they account for the secondary sexual character- istics in women such as growth of pubic hair, development of breasts and fat pads on the buttocks, and arresting of bone growth. Hence, it is no wonder that large doses of steroids taken orally are going to affect other tissues, organs, and functions in the body.

The scientific literature and popular reports are full of seemingly contradictory data linking the pill with blood clotting problems, increased blood pressure and various forms of cancer. Furthermore, if the pill, or other steroids, such as DES [diethylstibestrol] are taken during early pregnancy by women who do not know for certain if they are pregnant, abnormalities may occur in the embryo. In addition, if the pill is taken late in pregnancy, it can end up in the mother's milk and might cause complications in the suckling child.

There are some obvious explanations for this confusion and controversy. Much too little was known about the action of steroids and their analogs on humans when oral contraceptives and other steroids, such as DES, were first let out into the marketplace. Not enough testing was done, and significant information obtained from experiments on other animals was ignored.

Confusion was also occasioned by the fact that during this experimentation period, the dosage of steroids in the pills varied. Now the dosages are relatively low compared to those in earlier versions of the pill. Hence, many of the deleterious effects observed in women taking the pills years ago are not found in women taking the new lower-dose pills today. In fact, some reports indicate that rather than cause cancer, oral contraceptives actually prevent certain kinds of cancer and cyst formation.

Whatever you do, decide wisely, considering the latest information and advice. To start taking steroids orally is an important decision. You should first consult with a physician, discuss your family history, have a medical examination, and, if necessary, take certain tests. All these factors should help your doctor advise you on whether or not you can use steroids effectively as contraceptives with the benefits outweighing any possible risks. Proponents of the pill state that overall, the small risks resulting from oral contraceptives are trivial compared to the risks in bearing many children, especially to women in developing countries.

Assuming that you do take the pill, even proponents will warn you that smoking increases the chance of the deleterious side effects of the pill occurring. If, while on the pill, you experience chest, arm, or abdominal pain, shortness of breath, changes in vision, severe headaches, depression, jaundice, or other unusual reactions, report them to your physician immediately.

RU - 486

This French drug, which is taken after intercourse, may prevent pregnancy by a unique process. RU - 486 is a synthetic molecule that looks like the steroid progesterone, but does not act like progesterone. Nonetheless, RU - 486 can occupy receptor sites for progesterone that exist on cells lining the inside of the uterus, and thereby blocking progesterone from activating those receptors. When those receptors are blocked, the uterus will not develop its nutritive lining essential for the development of the embryo, and the fertilized egg will not implant and develop there. This drug has yet to be approved in the United States.

Non-Steroidal Pills

While some scientists are trying to make steroid oral contraceptives which have fewer or no adverse effects, others are attempting to develop oral contraceptives which contain no steroids. For example, an excellent oral contraceptive might be a form of inhibin or of the gonadotrophin releasing factor (GnRF) which would prevent ovulation in females or the production of sperm in males. Such hormones would be free of steroids and presumably would not produce the side effects associated with the use of steroid oral contraceptives.

Chapter 6

Menstrual Cycle and Female Hormones

Part II: A Second Look

By this time you may feel that you have read enough about the menstrual cycle, the hormones involved, their actions, their interactions, and their roles and helpful and harmful side effects when used as contraceptives. Yet we have hardly touched upon a variety of other important hormonal and physiological aspects of the menstrual cycle, such as the cause of menstrual pains (*dysmenorrhea*), the biological necessity of the monthly period, variations in the menstrual period, the failure to ovulate and menstruate (*amenorrhea*). I have delayed discussing most of these subjects until now because I believe that first you needed to comprehend the basic information contained in the preceding chapters.

To gain a full appreciation of the role that the menstrual cycle plays in our daily lives, we ought to be aware of why historically (and in some cultures yet today) women having menstrual periods have been subject to all sorts of social taboos, why they sometimes have been viewed as "cursed." There have been many superstitions related to menstruation. In some societies menstruating women were believed to bring on such disasters as the souring of wine, the breaking of violin strings, or even death. The ancient Hebrews considered menstruating women unclean and required them to be isolated and to refrain from preparing food for others. In the middle ages, the process of menstruation connoted sinfulness and inferiority of women, and menstruating women were not allowed in the church.

These taboos are not difficult to understand when we consider at least four facts: (a) In early times, there was wide-spread ignorance about the menstrual cycle, with no knowledge, of course, of gonadotrophins, steroids, and mechanisms of ovulation. (b) There were no widely available effective contraceptives then. (c) Marriages and sexual intercourse took place at an early age, usually soon after puberty. And (d) mothers nursed their children for long periods.

As a consequence, women menstruated infrequently during their lives because either: (a) They had not yet reached puberty. (b) Those reaching puberty soon were pregnant, and, hence, did not menstruate during the nine-month gestation period. (c) After delivering their children they suckled them for extended periods, sometimes for as long as two years; women who constantly suckle their children menstruate infrequently, if at all, because of the inhibiting action of prolactin on ovulation. (d) Because of the absence of contraceptives, most women were either pregnant or suckling their young most of the time. Those who did not die from childbirth or from other causes, usually bore from eight to twelve children each.

Assuming that females reached puberty at age twelve, and that each child was spaced about three years apart (say two years suckling the previous child and nine months gestation of the next child), then after ten children (thirty years) the women would be forty two years old. For the few that lived that long, after several more children they soon reached menopause and stopped menstruating completely. Thus, when considering a population of women in a community, "monthly" periods occurred quite rarely. And, when they did, most likely they could be correlated with some major or minor disaster which coincidentally occurred around the same time. Thus, can you now understand why a menstrual period might be considered a curse?

Also, this attitude may be related to the pains associated with menstrual cramps, or to the fact that a regularly menstruating woman is not bearing children, and hence, might be "cursed" with barrenness.

Menstrual Cramps

Up to this point in our discussion of the menstrual cycle, we have said little about the effect of the menstrual period on the overall behavior and physiology of women. Yes, male readers of this chapter, menstruation is usually accompanied by pain. Severe menstrual cramps, called *dysmenorrhea*, affect some thirty to fifty percent of women of childbearing age. A painless menstruation has been defined as a "figment of the male imagination."

Menstrual cramps may also be accompanied by fatigue, headaches, nausea, skin eruptions, and overall irritability. In rare cases, the sudden increase in the diameter of the blood vessels occurring prior to menstruation, causes some women to bleed from their nose or lungs. All of these responses vary greatly among individuals.

What causes the pain of menstrual cramps? It cannot be caused by irritation resulting from the sloughing off of the inner lining of the uterus, because there are no nerve endings in that lining. Current research appears to be giving us the answer. The major cause seems to result from a combination of two factors: (a) the increased contractions resulting from the overproduction of compounds called *prostaglandins*, and (b) the lack of oxygen to tissues that results indirectly from the prostaglandin-induced increase in uterine contractions.

Prostaglandin, so named because it was originally found in extracts of the prostate gland, has been known for years to cause uterine muscle to contract. Because of this fact, some procedures use prostaglandins to induce abortions. We now know that prostaglandins are released during labor and lead to labor pains. Thus, it would appear that menstrual cramps are like labor pains, only less severe. Prostaglandins not only cause the muscles of the uterus to contract; they can also affect the muscles of the stomach and intestines. The excess contraction of these muscles that occurs during menstruation may cause nausea, vomiting and diarrhea. Contractions of the muscles in blood vessels carrying blood to the brain can give rise to headaches.

All of the menstrual discomfort I have just described can be ascribed either directly or indirectly to prostaglandins causing muscle tissue to contract. Muscle contractions alone, however, are not sufficient to account for the severity of pain experienced during dysmenorrhea. Apparently, because of the increased prostaglandin-induced uterine contractions, sufficient blood and oxygen can not get to the uterine muscles fast enough. This lack of oxygen in contracting tissue causes a situation known in physiology as *ischemia*, a condition which leads to great pain. Ischemic pain can be experienced as muscle spasms or leg cramps. Ischemic pains also occur in heart attacks as a result of insufficient blood reaching the heart.

Menstrual uterine cramps, thus, can be considered a cascade of events beginning with the release of prostaglandins. These cause uterine muscle to contract as during labor. The increased contractions create a demand for oxygen that the available blood supply cannot fulfill. The resulting lack of sufficient oxygenated blood (ischemia) of the actively contracting tissue leads to a series of physiological changes causing severe pain of the type experienced in leg cramps or heart attacks.

Scientists are now designing new drugs which prevent body tissues from making prostaglandins, and hence prevent severe menstrual cramps and related symptoms. These drugs, packaged under a number of different names, such as ibuprofen, indomethacin, mefenamic acid, and naproxen, are known to have relatively few side effects. Doctors caution women having asthma or ulcers from taking these drugs, however.

Aspirin, a drug commonly used for relief from menstrual cramps, is now known to act by preventing tissues from making prostaglandins. For some unknown reasons, however, aspirin does not act as effectively in preventing prostaglandin formation in uterine tissue as it does elsewhere in the body.

Why Menstrual Periods?

We have learned that the menstrual cycle functions to provide a mature egg monthly, and a favorable environment for nurturing the embryo developing from the fertilized egg. But what purpose does the menstrual period, i.e. the monthly sloughing of the inner lining of the uterus, serve? Can't the egg be released and a favorable uterine environment be formed without the accompanying pain and inconvenience?

Another way to put these questions is to ask, "Does the menstrual period provide an evolutionary advantage to the human species?" After all, in biological terms, it is only the evolutionary advantages to the species that count in the long run. Of all mammals, the periodic monthly bleeding occurs only in humans, apes, and in old-world monkeys (those with long noses).

In place of a monthly cycle, other mammals (e.g. mice, rats, dogs, bears, seals, lions, etc.) go through an estrous period varying from about one to five times a year depending upon the species. In everyday speech, we

say that during its estrous period the animal is "in heat."

From my readings, I have found three different attempts to explain why it is advantageous for women to retain the monthly period, despite the pain and loss of tissue involved. These three views — a physiological, a developmental, and an anthropological one — all argue that the menstrual period provides an evolutionary advantage to the human species.

The developmental and physiological views of the need for the shedding process during the menstrual cycle go hand-in-hand. The developmental explanation, however, asks why the inner lining of the uterus evolved such a complex temporary array of blood-enriched spiral arteries not found elsewhere in the body. It suggests that these spiral arteries, in addition to supplying an environment for an embryo, are also there to provide large amounts of oxygen and blood. These eventually will be needed to support the development of the large brain found in primates.

The physiological perspective suggests that the inner lining of the uterus enriched with so many spiral arteries, cannot remain for too long a period in those nutritious conditions. If the inner lining does remain too long, it might become overripe, degenerate, and thus be unable to support the implantation of a fertilized egg that might come along at a later date. Hence, it must be shed and

reconstituted at frequent intervals, which happen to be monthly. And, a new fresh lining can be made again, once a month, which is ready exactly when a fertilized egg may be coming along.

The anthropological interpretation emphasizes the differences between those mammals having sporadic estrous periods versus primates having monthly cycles. Proponents of this view point out that usually female animals in estrous are receptive to having sexual intercourse only during that period. As an extreme example, the female rat copulates only on the single day that she releases an egg. Females of the human and some primate species, in contrast, are sexually active and can engage in sexual intercourse at any time during their menstrual cycle. (Some men and women avoid sexual intercourse while the woman is having her menstrual period, but this abstention is usually for esthetic rather than for biological reasons.)

Anthropologists explain it this way: If a woman can engage in sexual intercourse at any time, then she will be more attractive to her man. Consequently he will be at home more, will provide her with food, will protect her from predators and other men, and will do so for the long period necessary to raise her children until they become independent. It is this last reason which is important in evolutionary terms, to protect the young until they are independent. Indepen-

Compliance
dicto

dence in biological terms means to be able to raise one's own family. Thus, certain anthropologists conclude, by having monthly rather than yearly cycles, female primates can be active sexually all the time and keep the male around long enough to help raise the young.

Failure to Menstruate or Ovulate: Amenorrhea

There are a number of reasons why women fail to menstruate. Most deal with the oversecretion by the anterior pituitary of the hormone *prolactin*. As we have already mentioned, this hormone seems to act on the ovary by preventing it from ovulating. Consequently, if an oversecretion of prolactin prevents ovulation from taking place, then no corpus luteum will form. And if there is no corpus luteum to secrete high levels of progesterone and estrogen, then the inner lining of the uterus will not thicken and be shed as it would be during a normal monthly period.

Prolactin is normally secreted in women who are suckling their young because the prolactin is needed to stimulate the mammary glands to make milk. This oversecretion of prolactin in nursing mothers also keeps them from ovulating and menstruating.

When prolactin secretion is high in women who are not nursing, we have a syn-

drome known as *hyperprolactinemia*. Although this syndrome may account for thirty percent of menstrual problems, not much is known about its causes. Likewise, scientists do not know how prolactin prevents ovulation, nor why so many women have this syndrome. Hyperprolactinemia may account for the infrequency of menstruation in women who are physically active, are starving, or are suffering from anorexia nervosa (a mental condition in which women diet to the point of starvation).

A hopeful sign for the control of hyperprolactinemia is the discovery of a drug called *bromocryptine*. This drug prevents the secretion of prolactin by the anterior pituitary. Most women suffering from the overproduction of prolactin can return to their monthly cycle after taking bromocryptine daily for about two months. This drug is also used to prevent another effect of high prolactin secretion — the accumulation of excessive amounts of milk in the breasts, a condition known as galactorrhea.

Stimulating Ovulation

Some women cannot ovulate even though the level of prolactin in their blood is considered normal. Thus, bromocryptine, which acts by preventing prolactin secretion, will not help them. For these women other drugs or hormones are used. One problem, however, is that it is hard to regulate the doses of these drugs. That is why when the drugs were first used, we often read in the newspapers that women who previously could not have children, when put on these drugs, had sometimes as many as four or five children at a time. These are the same drugs used by doctors who surgically remove an egg(s) from women who must have children by in vitro fertilization techniques ("test tube babies").

Of the two compounds used, *clomiphine* (clomid) is the one of choice because it less often stimulates multiple ovulations (release of many eggs at once). The other substance used, called HMG (human menopausal gonadotrophin), often leads to multiple ovulations, sometimes overstimulating the ovaries so that they become enlarged.

HMG is extracted from the urine of women going through menopause. Their ovaries do not respond to the FSH and LH manufactured and released by their anterior pituitary. Consequently the ovaries of menopausal women do not ovulate frequently, do not form a corpus luteum, and do not produce large amounts of progesterone and estrogen. Accordingly, because these steroids are not released into the bloodstream, the production of FSH and LH is not inhibited, and the concentration of these two gonadotrophins increases greatly in the blood. These then get into the urine of menopausal women where they can be extracted and used by doctors to stimulate ovulations.

You may ask, "Why do doctors go through all of the trouble to extract FSH and LH (called collectively here HMG) from the urine of menopausal women?" Because FSH and LH are very expensive to synthesize in the laboratory, and it is much cheaper to extract those hormones from the urine of menopausal women.

Menopause (Climacteric)

Around the age of fifty, when the ovaries stop responding to FSH and LH, women begin the non-reproductive period of their lives known as *menopause* (climacteric). The over-all hormonal changes involved were described in the preceding paragraph. As far as the woman's body is concerned, the major hormonal change is the cessation of most steroid synthesis by the ovary. Thus, to avert some of the unpleasant changes that occur during early menopause, some women take estrogen pills, although not all physicians advocate such a therapy.

Because of this gradual decrease in steroids, a number of related changes occur during menopause. The uterus no longer goes through its monthly changes, and it becomes smaller. Its muscle is slowly replaced by non-contractile tissue. The tissue lining the vagina becomes thinner, fewer bacteria grow in vaginal fluids, and the acidity level in these fluids change. Vaginitis and vaginal dryness are common complaints of women going through menopause.

Changes in the steroids also affect the breasts, bones, skin, and circulatory system ("hot flashes"). Some of these changes are temporary. Others increase progressively with age. Better knowledge of how steroid hormones act may lead to alleviating some of the discomforts of menopause, but they certainly will not reverse it.

Psychological Barriers to Menstruation

It should be obvious by now that there is considerable interaction between the brain and endocrine hormonal system affecting the menstrual cycle. There is much that we know about these interactions, and even more that we do not. There are numerous cases of women who fail to menstruate because of some sort of psychological problem. There are two classic cases that come to mind, one of a woman who did not menstruate for fear of an unwanted pregnancy, and another who did not menstruate because she wanted to believe that she was pregnant.

In the first case, a young woman after her first sexual experience, was afraid she was pregnant and reported to a doctor a few weeks after she missed her first menstrual period. The doctor examined her and tested her urine for human chorionic gonad-

otrophin. He could find no evidence that she was pregnant and told her so. Nonetheless, she returned a number of times complaining that she still had not menstruated. The exasperated doctor, still finding no evidence of pregnancy, examined her once again, lied to her exclaiming, "There is a little blood; you are just starting to menstruate." The next day she phoned the doctor and said that she was menstruating.

The other story involves a woman in her late thirties who finally visited her physician for the first time after missing six or seven menstrual periods. She wanted children and was positive that she was pregnant. She told her doctor that she had experienced many of the signs of pregnancy, such as enlarged breasts and morning sickness. But most convincing was her enlarged abdomen. The doctor, however, found her uterus to be of normal size and that the large size was due not to a fetus, but to accumulated gas in her intestine. When the doctor convinced her that she was not pregnant, she passed an immense amount of gas, her abdomen returned to its near-normal size, and she menstruated a few days later.

Synchrony of Menstrual Cycles

Women, after living for long periods in close proximity to each other, as in college dormitories or in nunneries, frequently begin their menstrual cycles at very much the same times. No one factor has been singled out as responsible for synchronizing their cycles, for setting their biological clocks. Some believe that the factor is a substance or number of substances released in small amounts into the atmosphere by menstruating women. This gaseous substance(s) probably activates sensory receptors in other women living in the same area, and these receptors affect the brain-anterior pituitary complex so that the menstrual cycles start at the same time.

This idea is not as far fetched as it sounds. There are many examples in the animal kingdom of similar gaseous and usually odorless substances being released by one member of a species which affect the behavior and physiology of other animals of the same species. Such compounds go under the scientific name of *pheromones*. There most likely are a number of undiscovered human pheromones affecting our behavior.

A Side Effect of the Menstrual Cycle — Endometriosis

The trend for many couples to put off having children until the later years of marriage may lead to large numbers of women becoming infertile because of a disorder known as *endometriosis*. This illness occurs when bits of the inner lining of the uterus

(endometrium) which usually are shed on menstruation, imbed in tissues of the body cavity. Apparently during menstruation some bits of the inner lining become lost and back out into the oviduct and sometimes through it into the body cavity. Some of those bits can attach to the oviducts, ovaries, the outside of the uterus, and other places in the pelvic area. And, because these transplanted bits arise from the inner lining of the uterus, they still respond to estrogen and progesterone and thicken and bleed every month just as they would if they were in place inside the uterus.

An early sign of endometriosis is severe pain in the pelvic area just before the menstrual period. Another sign is irregular periods or pains during intercourse. The most serious of the symptoms, however, is sterility.

It is not known why endometriosis seems to be found more frequently in women in their thirties who have not yet been pregnant. The chances of experiencing endometriosis increase with the number of menstrual periods a woman has had in her life. Women who have had early pregnancies, however, are known to have fewer cases of endometriosis.

Medical scientists are working on several hormonal treatments for endometriosis. One of these lowers the response of the ovaries to gonadotrophins (i.e. FSH and LH). Consequently smaller amounts of steroids are released by the ovaries (see Fig. 5-2) and the bits of imbedded endometrium shrivel up and tend to disappear — at least for a while. Severe cases require surgery.

Rhythm Method of Contraception and Conception

Now you have the basic information about the menstrual cycle needed to understand the pros and cons of the rhythm method of contraception. In this method, the progress of the woman's monthly cycle is followed through various indicators, and the couple refrains from sexual intercourse just before and after the woman ovulates. The major problem with this method is the difficulty of predicting fairly exactly the day on which ovulation occurs. Among the indicators used are daily measurements of body temperature: It rises about a half a degree Fahrenheit when ovulation occurs.

In addition, because the thick cervical mucus becomes watery around the time of ovulation, some women using the rhythm method monitor the consistency of their mucus. Finally, some women experience a sharp abdominal pain at ovulation caused by the release of a little blood from the ovary into the body cavity.

A major fault with all these measurements is that beforehand you can only estimate the time of ovulation. You can be certain of it only after ovulation has occurred. Thus, if the woman has had intercourse a day or two before, the sperm will be in the oviduct waiting to fertilize the oncoming egg.

The rhythm method sometimes can be used effectively "in reverse" by individuals who wish to become parents and have had some difficulty in conceiving. By determining when the body temperature rises or when the cervical mucus becomes watery, a woman will have a fairly good idea of the next time that she will be ovulating. Thus, if the couple wishes to conceive a child, that is the best time for them to engage in sexual intercourse.

Chapter 7

Physiology of the Sexual Response

Nature has made sexual intercourse pleasurable as a sort of biological bribe to encourage reproduction and thus the survival and evolution of the species. As mentioned in the previous chapter, some anthropologists believe that women have monthly cycles rather than estrous periods a few times a year as do other mammals, to encourage men to engage in sexual intercourse with them throughout the year so that the men will stay home and provide for the family until the children reach maturity. Naturally, frequent sexual intercourse leads to more children, larger families, and greater responsibilities. Modern man has been the only animal species with the intelligence to develop contraceptives to circumvent the reproductive process for the sake of erotic pleasure — and, hopefully, for the building of a stable family unit.

The object of this chapter is not to enlighten you on the sociology, psychology, and practice of sexual intercourse. My object is to help you to understand the physiological basis of the sexual response, i.e. to give you an understanding of what happens in your body and that of your sexual partner during intercourse.

Much of what is known about sexual intercourse, however, still is mostly of a descriptive nature. By that I mean scientists can now tell us something about the contractions, expansions, blood flows, and rates of heartbeat taking place in our bodies during intercourse and about some of the nerve centers involved. They can not tell us much, however, about the neural control of sexual intercourse, or about why the various muscular contractions and expansions are necessary for the pleasurable sensations of orgasm. Other questions will occur to you as you read this chapter. Most likely science can not answer them — yet!

Why do we have so little scientific knowledge of this fundamental process which most people engage in with various degrees of frequency, and which is essential for the evolution and survival of the species? To answer that question, just ask yourself the following: How difficult would it be to get Congress, your church or synagogue, or the head of some large business to support and approve research on understanding the biological mechanisms involved in sexual intercourse? How would you study it? Would you use monkeys or humans for your experiments? If you used humans, whom would you use? How would you take your measurements?

Difficult questions to answer, are they not? Especially since society makes it difficult not only to do research on sexual intercourse, but even to talk rationally about it in public. Virtually everyone, including intellectuals, including readers of this book — and its author — snicker a little when talking about sexual intercourse, even during a serious discussion. Maybe a snicker or two is healthy to help us break the tension, so that we may discuss intelligently and maturely this important subject that affects our lives in so many ways.

Research Breakthrough

The first truly scientific research on sexual intercourse was conducted by Masters and Johnson in the mid 1960s. The way was paved for them, however, by Kinsey in the late thirties and in the forties. Kinsey, did not investigate the process of sexual intercourse, but published a highly reputable study on the sexual habits of American men and women. Although looking back today it is apparent that Kinsey's studies are limited because they covered primarily the habits of a cross section of middle class Americans at one time in our history, nonetheless publication of his results was revolutionary. Why? Because with publication of the "Kinsey Report," it became respectable, in some circles at least, to talk objectively about sexual intercourse.

Masters and Johnson

These scientists were the first to investigate seriously the physiological processes taking place during sexual intercourse. In their break-through research, they asked two fundamental questions: (a) What physiological changes occur in men and women as they respond to sexual stimulation and engage in sexual intercourse? (b) Are the response patterns that occur during sexual intercourse under voluntary, involuntary, or hormonal control?

As part of their research they recorded muscular, circulatory, respiratory, and sensory changes in the individuals involved. Their results are based on about ten thousand cases.

Masters and Johnson were very innovative and introduced many new techniques to carry out their studies. For example, (a) they measured the size, length and diameter of the male and female organs during different phases of sexual stimulation. (b) They used intrauterine electrodes to measure uterine contractions. (c) The subject's rate and level of heart contractions during sexual stimulation were measured using electrocardiograms. (d) Likewise, the respiration and blood pressure of the individuals were determined frequently.

The data were collected from subjects carrying out intercourse as well as from subjects while they were masturbating. In order to observe and record intravaginal responses, they used a specially constructed plastic penis fitted with a lens of a movie camera. They found that it did not make much difference which of these means for stimulating a sexual response was used: the reaction patterns given by males and females were independent of the method used.

Four Phases of the Sexual Response

Their overall findings showed that both females and males had similar sexual responses. Masters and Johnson, after measuring mostly circulatory, muscular and sensory changes, concluded that the human sexual response consists of four general phases: (a) *Excitement phase*, which includes the changes occurring during foreplay, i.e. before the penis enters the vagina. (b) *Plateau phase*, which is an intensification of changes occurring in the excitement phase occurring after entry and during the movements of sexual intercourse. (c) *Orgasmic phase*, which is the most intense of all phases. In males this response takes place mostly in the pelvic area and culminates with ejaculation. In females, orgasm is more of a whole body response involving numerous contractions of various muscles. (d) Finally, there is the *resolution phase* in which the body returns to the pre-intercourse condition in which there was no sexual stimulation. In males, the early part of the resolution phase, especially the time immediately after orgasm, is defined by Masters and Johnson as the refractory period. The refractory period is the time during which another erection and orgasm can not take place. The duration of the period can vary from minutes to hours depending upon the individual.

Myotonia and Vasocongestion

These two terms are applied to many of the muscular and vascular (blood) changes that occur during the sexual response. *Myotonia* refers to an increase in muscle tension. During sexual stimulation, many muscles of the body contract. Many of these contractions are not voluntary; they include such actions as facial grimaces, and twitching and clutching movements of the hands and feet (called *carpopedal spasms*). You should know that our muscles are usually never completely relaxed. They exhibit various degrees of contraction, even during sleep.

A change in the vascular system called *vasocongestion* refers specifically to an ac-

cumulation of blood in an area, such as occurs in the spongy bodies of the penis during an erection. An accumulation of blood in an area of the body also brings changes other than erections. These are discussed below.

Female Response

Excitement Phase

During the early stages of sexual stimulation, the heart rate and blood pressure start to increase. The nipples, clitoris and vagina begin to fill up with blood. Vasocongestion of the vagina causes it to release fluid. This so-called *sweating reaction* lubricates the vagina preparing it for sexual intercourse. Vasocongestion can also cause more blood to enter the skin of the breasts, abdomen and shoulders leading to a form of blushing called the *sex flush*.

During this phase muscles in many parts of the body show an increase in muscular tension (myotonia), and late in this phase the vagina lengthens and increases in width while the uterus rises until it pushes up toward the spinal column. In addition, the clitoris and other parts of the body become more sensitive, pleasurably so, to touch.

Plateau Phase

This phase can be considered as one in which there is a continuing and intensifying of all the changes that started to occur in the excitement phase. The heartbeat, blood pressure, and rate of breathing continue to increase, while the twitching and clutching movements of the hands and feet (carpopedal spasms) become quite noticeable. The pelvic region of the body starts to move involuntarily. The clitoris retracts, becoming extremely sensitive to touch.

By the latter part of the plateau phase, the body is considered to be in a delicate balance between the peak of excitement and orgasm.

Orgasm

Orgasm has been defined as the intense pleasurable sensations that take place during the sexual response when there is an "explosive discharge of accumulated neuromuscular tensions." The word orgasm is derived from the Greek *orgasmos*, which means "to swell, to be lustful." Orgasm in males is easily defined as those physiological changes which occur during ejaculation of the semen. In females, however, there is no such single indicator.

During orgasm in females, the lower vagina usually goes through a series of rhythmic contractions. The first of these

contractions last about 2-4 seconds each. These are followed by a series of shorter contractions that are less than a second apart. At the same time the uterus contracts, and sometimes the muscles closing the rectum also contract.

The blood pressure increases during orgasm, and the pulse may rise from seventy beats per minute to one hundred and eighty. Likewise, the breathing rate may increase from twelve to forty breaths a minute. The body becomes particularly flushed at this time.

Resolution Phase

After orgasm is completed, the body starts to return to its physiological state preceding erotic stimulation. The muscles relax and circulatory changes and breathing patterns return to normal. Compared to males, most females can still be stimulated erotically during the resolution phase and can be aroused to have multiple orgasms.

Male Response

Excitement Phase

In general, except for the different sexual organs involved, the responses of the male during this phase are similar to those of the female. This generalization is especially true of increases in muscular tension, blood pressure and rates of heart beating and breathing. The accumulation of blood in the penis leads to it becoming erect. The erections can last from half an hour to an hour, usually depending on age and health of the individual. During this phase the Cowper's gland releases a few drops ("love drops") of its secretion along with a little semen. The penis is also more sensitive to touch.

Plateau Phase

During this phase, the tip of the penis swells even more, and a sex flush becomes more pronounced. The heart rate, rate of breathing, and blood pressure stay at the same high level as during the excitement phase, or increase even more. Muscular tension increases and carpopedal spasms may take place.

Orgasm

Although there is a general loss of muscular control in males during orgasm as occurs in females, and an increase in heart rate and flushing, the main event of orgasm in males revolve around ejaculation. The testes swell, and the muscular elements of the epididymis, seminal vesicle, prostate gland, and vas deferens compress the semen behind a circular muscular valve *(sphincter)* situated at the beginning of the ejaculatory duct. At this time the male senses a "feeling of inevitability" indicating he can not hold back the flow of semen. The sphincter valve then opens and the semen is pumped out of the penis through the ejaculatory duct and urethra by a series of pulsating pumping movements. Once the valve opens, another sphincter valve linking the bladder to the urethra closes tightly so that urine will not mix with the semen. The intensity of the orgasm varies with the individual.

Resolution Phase

As in females, during this phase the circulatory and muscular systems return to the individual's state before sexual stimulation. In males, the early part of the resolution phase is called the *refractory period*. During the refractory period, which may last from about ten minutes to hours, rearousal to orgasm is impossible. Many factors may affect the length of this period. Only about six to eight percent of men are said to have more than one orgasm during a sexual experience.

Graphic Comparison of Response by Men and Women

Masters and Johnson summarized the major aspects of the sexual responses by males and females in two graphs, which I present in a modified form in Figures 7-1 and 7-2. In both figures the vertical axis represents "the level of sensory, vascular, and muscular tensions," whereas the horizontal axis represents the relative length of time of the sexual response. In males (Fig. 7-1), the plateau level, which can be reached rather rapidly, can last as short as one to two minutes before orgasm takes place, or in some cases the plateau phase can be extended for as long as an hour. Orgasm is immediately followed by the refractory period of the resolution phase, during which no erection or second orgasm can take place. In some men, as indicated by the dashed line, a second

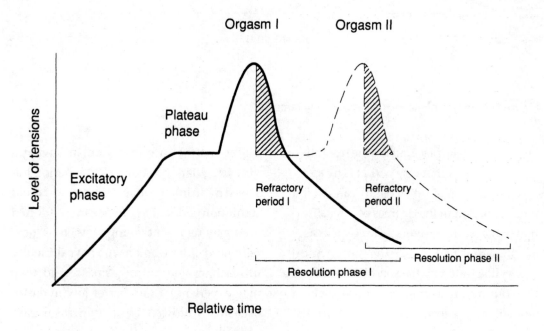

Figure 7-1. Stages of the sexual response in males.

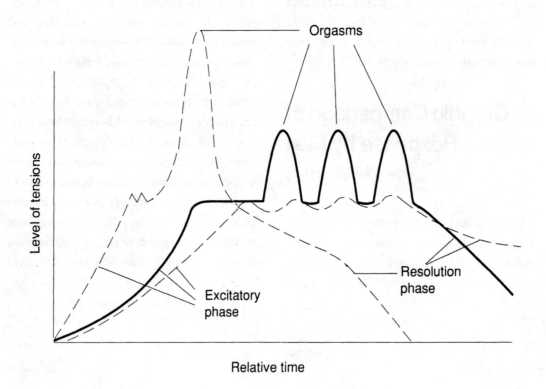

Figure 7-2 Stages of the sexual response in females.

orgasm may take place later in the resolution phase after the refractory period is over.

Females, on the other hand, can exhibit more variation in the degrees of intensification of the sexual response. Three such cases are shown in Figure 7-2. The more typical case is illustrated by the solid line. It shows that during an average response cycle, a woman may experience one or more orgasms. A second example, the dashed line,

represents the case of a woman having a very irregular and short plateau period followed by an intense long orgasm and a rapid resolution period. The third example (dotted line) may represent young women or inexperienced older women who have difficulty in reaching one full orgasm. Instead they may experience a number of low intensity orgasms, followed by a long resolution phase.

Some Generalizations

Although research dealing with understanding the nature of the sexual response is still in its early phases, we nonetheless can draw some generalizations. First, except for the process of ejaculation in males, and multiple orgasms in females, the sexual responses of males and females are relatively similar in their physiological characteristics.

Secondly, virtually all, but not all, aspects of sexual intercourse and behavior are similar in human beings and other mammals. In human beings, however, much of that behavior is under cerebral (voluntary) control. In other animals, by contrast, we say that their sexual behavior is "instinctive." By instinctive, we mean that those behaviors are controlled genetically by involuntary neural activities that may be synchronized with the season and activated by pheromones that are released by females who are ready to ovulate. Most likely as human beings evolved some of the instinctive behaviors came under voluntary control.

As these evolutionary changes in the brain of human beings were taking place, changes also occurred in those brain activities affecting the release of the hormones causing ovulation. Whereas most mammals release their eggs a few times a year during their estrous periods, humans do this at monthly intervals throughout the year. Hence, with humans sexual intercourse and fertilization can take place at almost any time, whereas in most mammals both processes are regulated to take place only during specific breeding seasons.

Finally, erogenous sensations appear to act as a sort of biological bribe to encourage sexual reproduction, and, thereby, survival of the species. Although ejaculation occurs in other mammals, it is not known if in other mammalian species the females experience orgasm as do human females. We do know that through the use of contraceptives, however, only human beings are able to bypass fertilization and still enjoy the erogenous pleasures of the sexual response.

Future

There is still much we do not understand about the sexual response and associated behaviors. In addition to learning more about the neurological and physiological processes affecting the response, there is a great deal to learn about psychological mechanisms affecting such behaviors as heterosexuality and homosexuality, aberrant sexual behaviors, and the effects of pornography and violence on sex. The future of our understanding of all these behaviors will depend on further research on these subjects.

CHAPTER 8

Venereal Diseases

Venereal diseases are those which are transmitted during intimate physical contact such as occurs in sexual intercourse. I include a section on venereal diseases in this book not only because they can be one of the consequences of sexual intercourse and because venereal diseases constitute a serious public health problem, but most especially because the emphasis of this book is healthy pregnancies. It is well known that the complications of such widespread venereal diseases as gonorrhea, syphilis, and herpes are magnified in pregnant woman and can be extremely damaging to the child being delivered. With the HIV virus, which can lead to AIDS, infection of the fetus will lead to the eventual death of the infant.

Venereal diseases are named for Venus, the goddess of love. Because we know that many cases of venereal diseases are transmitted in sexual acts devoid of love, many scientists refer to them as STDs, i.e. sexually transmitted diseases. Some societies tend to refer to venereal diseases as those that primarily affect their neighbors. For example, in France, they are referred to as the Spanish sickness or the Neapolitan disease. Of course in Spain it is called the French disease. The custom is not much different in Germany, where venereal disease is called the Polish pox, or in Poland, the German pox.

Regardless of what name they go by, venereal diseases are major world health problems. For example there are approximately two hundred million cases of gonorrhea worldwide with over two million new cases each year, and fifty million cases of syphilis with over one hundred thousand new cases being reported each year in the United States alone. In 1977, there were over three million cases of venereal diseases in the United States, with most being gonorrhea; even so, in the United States one thousand individuals die monthly just from syphilis alone. Supposedly because of the sexual revolution that has taken place in recent years, a "new" venereal disease is spreading in near epidemic proportions. This disease, genital herpes, has been reported to affect from five million to twenty million Americans. Because it is a disease transmitted by a virus, it is not treatable with antibiotics such as penicillin, as are gonorrhea and syphilis. Hence, once a person is infected with genital herpes, he or she is usually infected for life.

With the AIDS virus the problem is far worse. At this stage of our knowledge of AIDS, once the virus expresses itself, the infected individual has no chance of recovery and is doomed to an early death.

Venereal diseases are not the exclusive miseries of the poor and of prostitutes. To the contrary, they affect individuals of all economic, social, and educational levels. Just to name a few well known individuals known to have contracted venereal diseases, we can mention the artists Van Gogh and Goya; the composers Beethoven and Schubert; the authors Keats, Thackeray, Moliere, Boswell, Oscar Wilde, Jonathan Swift, Strindberg, Walt Whitman; the philosophers Nietzche and Schopenhauer; the clerics Cardinal Richelieu, Pope Alexander VI, and Pope Leo X; the heads of state Napoleon, Ivan the Terrible of Russia, Louis XIV of France, and Henry VIII of England; and then there was Casanova. Today, with the availability of antibiotics, cases of notable public figures who have at one time or another contracted venereal diseases other than AIDS are generally less well known.

Kinds Of Venereal Diseases

There are over a dozen diseases known to be transmitted through sexual intercourse. Although they have a similar means of transmission, the various venereal diseases differ both in the type of infecting agent, and in the symptoms of the diseases. Some are bacterial in origin (e.g. gonorrhea and syphilis), whereas others are caused by protozoa (e.g. Trichomonas), viruses (herpes and AIDS) or by other agents.

Gonorrhea

Often called "clap" after the Clapian district of Paris where many prostitutes lived, gonorrhea is caused by bacteria which infect the moist membranes of the vagina, penis, anus, mouth and throat. If left untreated, the bacteria can pass up the genital tract and infect it to cause sterility. In a small percentage of the patients, the bacteria can enter the bloodstream to spread throughout the body to cause arthritis, heart disease, and other ailments.

In females, gonorrhea can be extremely dangerous because the disease is usually advanced before it is detected. Most infections do not give any obvious detectable early symptoms. As a rule infections are first noticed when a yellow green pus is discharged from the cervix. Sometimes a burning sensation is experienced during urination, and around the vulva. Frequently a woman first becomes aware that she is ill when her sexual partner reports that he is infected.

Once the infection starts, it can reach the oviducts within two to three months. The pus which forms there can leak into the uterus, onto the ovaries, and into the abdominal cavity. Infection of the oviduct and adjoining areas can lead to pain, such as in backaches, and, in twenty five percent of the cases, to sterility. The sterility is usually caused by the oviducts becoming permanently blocked by scar tissue so that neither eggs nor sperm can pass through them.

Whereas about forty percent of women having one sexual act with an infected partner can become infected, some reports indicate there is one hundred percent chance of becoming infected if the woman is simultaneously using steroid oral contraceptives. Apparently, the pill affects the moist environment of the vagina so that it is more receptive to the bacteria causing gonorrhea.

In males, the infection of the urethra of the penis can usually be detected within two weeks of exposure to the bacteria. There is usually a pussy discharge from the penis, a burning sensation there, and the urine appears cloudy and occasionally shows traces of blood. During the third and fourth weeks following infection the bacteria spreads deeper into the genital tract, although at the same time the initial symptoms seem to disappear. The prostate gland, once infected, becomes larger and presses against the bladder and rectum, making elimination of body wastes painful. Infection can also spread into the vas deferens and into the epididymis leading to epididymitis, which causes a hard painful swelling along the testes. Scar tissue resulting from these infections of the genital tract can lead to sterility by preventing the passage of sperm from the testes. Relatively speaking, however, the effects of gonorrhea in males, compared to those caused by syphilis, are less devastating and last for shorter periods.

In addition to being transmitted by sexual intercourse, gonorrhea can be transmitted from one individual to another by hands that may be covered with smears of pus laden with bacteria. Because the eyes of newborn babies may become infected when passing through an infected vagina, it is now customary to place drops of silver nitrate or an antibiotic in the babies' eyes as soon as they are born to prevent blindness from developing.

Treatment of individuals with gonorrhea is usually successful when penicillin is used as the antibiotic. In recent times, however, many penicillin-resistant strains of gonorrhea have developed through mutation. Thus, whereas once only a dose of two thousand units of penicillin was sufficient to cure someone of gonorrhea, now at least 4.8 million units are required. There are a number of new strains of penicillin-resistant gonorrhea from southeast Asia that are currently spreading in the United States. Because of the emergence of these penicillin-resistant strains, some new antibiotics have been developed. Among the newest is spectinomycin. Although expensive, spectinomycin is relatively effective against the Philippine variety of gonorrhea. Unfortunately, most likely a spectinomycin-resistant strain of gonorrhea will develop in the future.

Syphilis

The name of this disease is derived from the legend of the shepherd, Syphilis, who angered the sun god, and who was subsequently the first victim of a plague sent to earth by the angered god. This sexually transmitted disease affects one hundred thousand new individuals per year in the United States, and about three hundred to four hundred newborn babies here contract the disease from their infected mothers.

Syphilis is caused by bacteria. If not cured early, this disease can go through four stages. *Primary syphilis* is usually noticed ten through twenty-eight days after infection when a hard and painful sore, called a chancre, develops on the vulva, deep in the vagina, at the rectum, on the shaft of the penis, or in the urethra of the penis. The infectious spirochete causing syphilis can be detected by a blood test. Although the chancre may disappear within several weeks, the infection may persist in this stage for six months to a year. It is important to treat the disease early after the infection is detected. It can be successfully eliminated in this stage as well as in the next phase, by a single strong injection of penicillin.

Secondary syphilis develops as late as one year after the first chancre appears. Individuals during this second stage are highly contagious and exhibit many symptoms including varying degrees of skin rashes and lesions, temporary baldness, fever, and swollen glands. Almost any organ of the body may be infected. The infection during this stage normally can be eliminated by a high dosage of penicillin.

During the third stage, or *latent period*, the signs and symptoms of the disease disappear, but the infective spirochetes persist in some organs, such as the brain, for as long as twenty years or more. Nonetheless, the individual is usually not contagious during this period, except that a pregnant woman may infect the child she is carrying.

The *last stage of syphilis*, which may affect twenty-five to thirty percent of untreated individuals, begins ten to twenty years after the latent period. This last stage is the most dangerous for the infected individuals. If untreated, they become disabled by damage to the heart and/or to the brain and spinal cord. Many of the victims become blind, insane and/or crippled. They often exhibit large disfiguring abscesses called *gummas.*

As mentioned, in addition to syphilis spreading to adults through sexual intercourse, the disease can be transmitted to the fetus from an infected mother. We call the disease contracted this way, *congenital*

syphilis. If the fetus is exposed to the spirochete in the fourth month of pregnancy, it may die or be born disfigured. If the fetus becomes infected late in pregnancy, it may be born showing symptoms of the disease, and may become disabled when the disease has gone through all its stages. The spread of the disease in the fetus can be stopped by treating the mother with penicillin during pregnancy, especially during the first months.

Although antibiotics, especially penicillin, can cure congenital, primary, and secondary syphilis with as little as a single injection, a person, once cured, can be reinfected over and over again. Furthermore, as with gonorrhea, the infectious agent may gradually evolve a resistance to some antibiotics.

Genital Herpes, or Herpes Simplex Virus, Type II

Genital herpes is now recognized as among the more common venereal diseases in the United States after gonorrhea. Unlike gonorrhea, however, herpes infections can not be eliminated by treatment with antibiotics. In fact, there is no sure cure that is specific for herpes. In 1982, reports of genital herpes in the United States ranged from five to twenty million new cases being diagnosed each year. There are over one hundred thousand people infected with herpes in Dallas, Texas alone. Some public health officials ascribe the rapid increase in genital herpes to the new sexual freedom associated with certain contemporary life styles, especially among educated younger people.

To avoid confusion, genital herpes should be distinguished from other types of herpes. One kind, Herpes zoster, is known to cause chicken pox in children and "shingles" in adults. The most common form of herpes, called Herpes simplex I is also known as "above-the-waist" herpes; this virus causes the so-called fever blisters or cold sores that usually are found on the lips. Although this kind of herpes is not the major cause of genital herpes, about five percent of genital herpes infections can be ascribed to it.

We are concerned here with Herpes simplex II, the kind that occurs "below-the-waist" and that first infects the genitals, thighs, and buttocks. Infections with genital herpes may go through several stages. First the disease may go unnoticed during an incubation period of a few days up to three weeks. During that time the affected areas may tingle or burn somewhat. In females, some blisters may develop on the internal walls of the vagina. Blisters become very common during the next stage of the infection, and they may be found on the penis, vulva, and thighs. The blisters eventually rupture, leaving painful ulcerated tissue in the infected areas, and then heal within two weeks. In women, the ruptured blisters in the lips of the vagina can be especially painful during urination.

The blisters may reappear at varying frequencies without the infected individual having any more sexual contact with infected partners. The exact causes of the reoccurrences are unknown. Some may be due to psychological stresses, or to such physiological stresses as caused by menstruation, steroid oral contraceptives, colds, or heat. During the latent periods when blisters do not appear, the virus may remain undetected and dormant in nerve cells of the infected individual. The number of new eruptions gradually declines with time; for example, a person may have five attacks in the first year of infection, one to three in the second year, and even fewer thereafter.

Perhaps two of the most distressing side effects of genital herpes are its effects on newborn infants of infected mothers, and the possibility that herpes may increase the incidence of cervical cancer. As regards herpes affecting newborn infants, its effects are threefold. For one, herpes is particularly dangerous because it is a virus and therefore can pass from the mother, through the placenta, and into the tissues of the developing fetus. It is estimated that one fourth of all newborn babies of mothers who were infected during pregnancy may have congenital malformations. Of these, one half die and the rest may have serious brain damage. Secondly, in those cases where the herpes virus did not pass through the placenta, there is the possibility that the child may contract the virus when it passes through the infected vagina. Hence, Cesarian deliveries are usually recommended for pregnant women known to be infected with herpes. And, thirdly, there is increasing evidence that women infected with herpes have three times the miscarriage rate as uninfected women.

As regards cervical cancer, there is a strong suspicion that at least one form of the cancer is caused by an infectious agent, most likely genital herpes. For example, cervical cancer is found at higher rates in prostitutes, women who have multiple sex partners, and in women who have married men whose previous wives had cervical cancer. But since the Pap test allows cervical cancer to be detected early in its development, there is no major concern of the cancer spreading if a hysterectomy is performed early enough.

There is no immediate cure for herpes, although a number of ointments have been developed to ease the pain of the ulcerated blisters. Herpes cannot be killed by antibiotics, as can the microorganisms causing gonorrhea and syphilis, because the latter are bacteria, whereas herpes is a virus.

What is a virus? A virus may be described as a group of protected genes that can penetrate a cell and force that cell to divert its energies to manufacture more of that virus. For this reason, once a virus such as herpes gets into the host cell, the number of virus particles increases rapidly. Since viruses are made up mostly of the genetic material DNA (see Chapt. 16), if we apply any drug which prevents the formation of the DNA of herpes, we may also interfere with the normal operations of the host cell which are also controlled by DNA. Nonetheless, research continues on this dehabilitating disease, and some new ointments show promise of shortening the initial bouts with a herpes infection.

Because of the near epidemic spread of genital herpes, individuals who have contracted the disease have many psychological problems. Aside from the mere stigma alone, they fear having sexual intercourse, getting married, and getting pregnant.

AIDS

Acquired Immune Deficiency Syndrome (AIDS) is the name applied to a deadly venereal disease, first recognized in the early 1980s. If unchecked, AIDS has the potential of spreading in epidemic proportions. The frightening aspect of this disease is that no cure yet exists; once symptoms of the disease show, death within a few years seems certain. Consequently, many articles and books on this subject are being written. The purpose of this section on AIDS is to present to you in simple terms the present state of knowledge regarding this disease. With such information you should be able to evaluate and understand articles and books that you may encounter over the next few years. You should also be able to make intelligent decisions regarding adapting your lifestyle so that you will avoid contracting AIDS.

How Does it Spread?

There are three ways that AIDS is spread, and all have to do in some way with blood. AIDS is most commonly spread through *sexual intercourse* because the AIDS virus (HIV) is present in semen. But in order for the virus to infect, it has to enter the blood steam of the recipient. The chances of infection are increased if the genital tract has open sores or cuts. Thus individuals having sores from another venereal disease will be more likely candidates for contracting AIDS. So will individuals who practice forms of sexual intercourse that lead to open cuts or tears, such as might occur during anal intercourse.

Since the disease can be contracted if the virus can get into the bloodstream, it is easy to understand why AIDS is spreading so fast among drug addicts who share the same needle for injecting drugs into the bloodstream. At one time, before screening of blood donors for AIDS became widespread, the virus was sometimes transmitted through transfusions carried out with infected blood.

A third way for the virus to be transmitted is for it to pass from the blood of an infected mother through the placenta to the blood of her fetus. As you will learn in Chapter 22, viruses are among the agents that can pass readily through the placental barrier. Thus, one of the cruel facts of this disease is that children born to mothers, usually drug addicts, who have the AIDS virus, are doomed to an early death.

The evidence to date is that the virus is not transmitted through the saliva, sharing cups, kissing, casual contact, and hugging.

Possible Scenario

Those who have contracted the virus may not show symptoms of the disease for five or more years. During that time they may pass the virus along to many others, depending upon their sexual and drug habits. Once the symptoms of the disease appear, death usually occurs in a little more than one year. Unlike the situation with most other viral diseases, specific antibodies do not build up to the AIDS virus because the virus keeps on mutating and changing the nature of its protective protein covering.

Where Does the Virus Attack and How Does it Kill?

How AIDS kills!

The virus infects and destroys the white blood cells of the immune system known as T-4 lymphocytes. The white blood cells are the body's means for fighting infections. They act by capturing the infectious agents, such as bacteria or viruses, and digesting them. Now you can see why AIDS is such a deadly disease; Once those white blood cells are destroyed, the body can no longer fight off infections. Hence, it is not the AIDS virus per se that kills the individual. It is other diseases that kill, such as pneumonia, tuberculosis, meningitis, a rare form of cancer called Kaposi's sarcoma, and a protozoan disease known as toxoplasmosis. Individuals with a healthy immune system and healthy white blood cells usually can fight off those diseases.

The virus also infects certain brain cells, and thereby causes *mental retardation* in newborns and *dementia* in adults. By dementia we mean that infected individuals lose their ability to think; they become confused and apathetic, and eventually become mute, incontinent, and blind.

AIDS Related Complex (ARC)

Not everyone who becomes infected with the virus, however, gets the disease. The U.S. Public Health Service predicts that over two millions Americans have already contracted the virus, and that by 1991 more than a quarter million Americans will have had the disease and will die from it.

It is possible to have contracted the virus, but not show any of the symptoms of AIDS other than the production of antibodies. Some of those harboring the virus my develop what is known as the *AIDS Related Complex* (ARC). Such individuals may have a number of the effects of AIDS, such as fever, diarrhea, weight loss, fatigue, and swollen lymph glands, but they do not show the symptoms of the deadly form of the disease. Finally, those who contract the deadly form of the disease usually die of the factors described above, factors usually controlled by a healthy immune system.

How Can an Infection be Detected?

To date there is no reliable test for the AIDS virus. Instead, there are a number of tests for antibodies that form in the blood of infected individuals in response to the AIDS virus. The antibody test is not always perfect; sometimes a false positive response is given by individuals who do not have the virus, but who may be suffering from some other ailment, such as malaria.

Furthermore, even though the antibody to the AIDS virus can be detected, it takes about two to ten weeks or even longer for an infected individual to produce sufficient antibody for it to be picked up by most tests commonly available today. Thus, for example, a person who has been infected with the virus for about a week, will test negative for the antibody, will have the impression that he or she does not have the virus, and may continue sexual and/or drug practices that may transmit the virus to other individuals.

Another difficulty of the current tests is that although they will tell you whether or not you are producing antibodies to the virus, they will not tell you whether or not you will ever develop the deadly form of the disease.

Are there any Cures?

AIDS is a virus and, like the virus Herpes, is not killed by antibiotics as are bacteria. Because the AIDS virus is constantly mutating and changing its protective protein coat, chances of developing a vaccine to it in the near future appear very slim. There are a few drugs that slow down the progress of some stages of the disease, but they are not cures.

AIDS is a very complex disease, and although there is much excellent research going on right now attempting to understand it, medical researchers do not expect any miraculous cures in the very near future. One of the unusual biochemical steps involved in the reproduction of the AIDS virus is described in Chapter 16.

Ways to Halt the Spread of Aids

Until a vaccine and a cure are discovered, about the only way to halt the spread of AIDS is by public education and by peer pressure. It is important that people understand AIDS, know how it is transmitted and detected, and know how to protect themselves. Although it is difficult to change people's behaviors, it is possible to modify them.

There are ten basic rules to follow:
1. Stay in a mutually faithful relationship with your sexual partner in which neither of you use intravenous drugs.
2. Use condoms correctly to prevent exposure to semen or to blood.
3. If you are starting a new relationship, make certain that you learn the previous history of that individual.
4. Remember that most people today who are harboring the virus have had no symptoms of the disease and do not know that they are infected.
5. If you think you are infected, tell your partner.

6. If you think that you are infected, get tested.

7. Tell all of your friends, your children, that AIDS is a new disease, that it kills, and that there is no cure for it.

8. If you already are taking drugs intravenously, do not share needles or syringes with anyone else.

9. If a friend or family member contracts the deadly form of AIDS, care for that person. You cannot get infected except through the means described in this Chapter.

10. If you become pregnant and think that you are infected, consult with your physician immediately.

Chlamydia

Chlamydia is an unusual bacteria that is spread sexually and that infects millions of Americans each year. It ranks with gonorrhea and genital herpes as one of the more common venereal diseases in this country. It strikes both sexes, has no early symptoms, and is difficult to diagnose. In females it can lead to a scarred oviduct, infertility, and ectopic pregnancies. The disease can be treated with the drug tetracycline.

The outward symptoms are rather general: burning and painful urination; vaginal discharge; pelvic and abdominal pains similar to those experienced by individuals infected with gonorrhea; and symptoms like those observed in pelvic inflammatory disease.

Chlamydia can survive such harsh conditions as freezing, even temperatures as low as -70° Centigrade, and is known to have been transmitted with sperm that has been stored in sperm banks.

Other Venereal and Related Diseases

Some of the lesser known venereal diseases can also cause significant problems. We have already mentioned that yeasts can infect the genital area. One of them, called *Candidiasis*, can lead to a yeast growing in the mouth, vagina, or intestine of almost fifty percent of healthy individuals. Heavy infections can cause a severe vaginitis. Another type of vaginitis, *trichomoniasis*, can be caused by infections by the protozoan parasite *Trichomonas vaginalis*. Although not a serious disease, trichomoniasis can make sexual intercourse painful. This disease is treatable with certain drugs.

Venereal (genital) warts are caused by a virus (*human papilloma virus, HPV*) that proliferates easily in such moist areas as the penis shaft and under the foreskin in uncircumcised men, around the vulva, vaginal wall, and cervix in women, and sometimes around the anus. Even though the infection rate of HPV has been thought to be relatively low compared to that of herpes, in recent years the incidence of the disease has been on the increase, especially among college students. Condoms give some protection against the spread of the virus, but, unlike the AIDS virus (HIV), the virus causing genital warts can be transmitted by sexual skin-to-skin contact in areas near the infection. Some strains of the virus lead to cancer of the cervix, penis, or anus. HPV may be passed on to the fetus by infected pregnant women. In severe cases the warts may partially obstruct the vagina during delivery.

Also of note are two conditions caused by *lice* and *mites*. One is *genital pediculosis*, also called "crabs." This condition usually noticed because of itching in the pubic area, is caused by a blood-sucking louse which is transferred during sexual contact. The other condition, known as *scabies*, is caused by a mite which deposits its eggs into the skin. The skin becomes irritated and itchy. The resulting itch can become extremely intense. Scabies can be transmitted either nonsexually or through sexual contact. Both crabs and scabies are treatable by the application of solutions of certain drugs to the skin.

Controlling Venereal Diseases

Obviously the best way to prevent the spread of venereal disease is to abstain from intimate sexual contact with others. Without trying to be facetious, this method, i.e. abstention, works with celibate individuals. Along these lines, the recent scare of epidemic incurable genital herpes has led to a decrease nationally in the spread of gonorrhea.

If a male choses to live a noncelibate life, however, the safest way for him to avoid contracting a venereal disease is to use a condom during sexual intercourse. Likewise, if a female is unsure of the health of her partner, she should insist that he use a condom during intercourse to lower the chances of his transmitting the disease. It would also help if bactericidal ointments were used simultaneously.

If, however, despite all precautions, an individual believes that he or she has contracted a venereal disease, then it is critical to see a personal physician or attend a VD clinic for proper diagnosis and treatment. If an infected person is having sexual contact with one particular partner frequently, then obviously it will do no good to be treated alone. Why? Because, even though once cured, that same person could pick up the disease again from the partner who most likely is also infected. As the poster in the VD clinic reads, "Love is going to the VD clinic together." By so doing, individuals can avoid the *"ping-pong effect,"* i.e. the transmission of the disease back and forth between partners who are treated at different times.

In reality the situation looks quite bleak for preventing the spread of venereal diseases before they reach worldwide epidemic proportions. Some diseases, like AIDS and herpes, are presently virtually untreatable; others, like gonorrhea, may develop resistance to antibiotics. What we need is a huge influx of worldwide governmental funding to promote research dealing with the eradication of venereal diseases, and to educate the masses.

The education, however, should not be restricted solely to warning individuals of the consequences of venereal diseases and of the effectiveness and availability of condoms. The education should also be aimed at individuals of puritanical upbringing who consider even public discussion of such subjects as venereal diseases as sinful. The research and education go hand and hand. After all, how will societies pass regulations supporting research if the public will not even discuss the subject and if it believes it is God's way of punishing lascivious people? Can you imagine, today for example, church leaders and the Parent Teachers Association recommending that all students be vaccinated against the clap and herpes as they are for small pox? It may eventually get down to that once the gonorrhea bacteria becomes resistant to antibiotics and if no reliable cure is found for herpes.

Ecology of Venereal Diseases

Aside from learning about the debilitating effects of venereal diseases, and how to avoid or cure them, these diseases can tell us much about how our bodies interact with foreign microorganisms in general. For example, many microorganisms are helpful to the body, especially those which live in our intestines. Some are important in helping us to digest and assimilate food. Others in the intestine actually provide our bodies with essential vitamins that we may not get from other sources. The microorganisms live deep inside our intestinal tract and get there through the food we eat.

Venereal microorganisms cannot survive in the two major ecological zones of the surface of the body: the dry skin, where few organisms survive, and the temporarily moist areas, such as the arm pit and groin areas. Those latter areas, which retain moisture, allow many bacteria to grow there. Those bacteria are harmless, and merely are annoying because of smells that they give off. As a result, there is a multibillion dollar business devoted to neutralizing or masking those smells.

The favorable ecological zones for the venereal organisms to establish themselves, however, are the *transitional* areas intermediate between the outer surface of the body and the organs and passageways deep within our bodies. The transitional areas, which require intimate physical contact between individuals for microorganisms to be transmitted, include the nasal passages, mouth, and anus; in men they also include the urethra and area under the foreskin, and in women, the vagina.

It is noteworthy that non-venereal microorganisms may also inhabit those areas. For example one third of pregnant women, and one fifth of women who are not pregnant, become infected with a yeast that causes a severe itching of the vagina and a thick white sticky discharge. This infection is also more prevalent in women on steroid oral contraceptives because of the high levels of progesterone which favor growth of this strain of yeast. When detected, these yeasts can usually be kept at a level that no longer causes unpleasant symptoms. Ironically, because yeast infections can be partially detected by the odor they emit, women who use feminine hygiene deodorant sprays may unintentionally postpone the diagnosis and treatment of those infections.

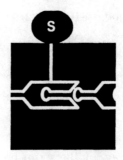

CHAPTER 9

Introduction to Gamete Formation, Genetics, and Molecular Biology

Why Study the Formation of Sperm and Eggs?

If we refer to the material covered thus far in this book as "sexy," then we can refer to the material covered in this and the next seven chapters as those describing the reason for sex. As we learned in Chapter 1, the major reason for reproducing sexually instead of without sex, i. e. asexually, is to provide genetic variation as a means for evolution, and, therefore, for survival of the species. To allow for genetic variation, we develop gonads to produce sperm and eggs. Because of the importance of sperm and eggs in sexual reproduction, I will devote this chapter and the next two to describing the development and structure of the gametes.

Knowledge of the processes leading to the formation of sperm and eggs will also help us to understand one of the major causes of birth defects and natural abortions, that is abnormalities in the structure and number of our chromosomes. These abnormalities are given the inclusive name of chromosomal aberrancies.

Here are some startling statistics. About one half of all human eggs that are fertilized never develop into infants. Either these fertilized eggs do not implant into the inner lining of the uterus, or, if they do implant, they abort naturally early during pregnancy — often without the mother even knowing that she was pregnant. From one third to one half of those fertilized eggs that either never implanted, or aborted early, are believed to have chromosomal aberrancies.

What exactly do we mean by *chromosomal aberrancies*? The term refers to some abnormality in the number and/or structure of the 46 chromosomes that are found in the nucleus of our cells. One aberrancy, for example, is the appearance of an extra chromosome (number 21) in the cells of individuals having Down's syndrome. Another aberrancy can result from the lack of a chromosome, such as occurs in individuals having Turner's syndrome (Chapt. 12). Still another aberrancy can occur if a piece of a chromosome is missing (Chapt. 12).

Such chromosomal aberrancies usually result from natural causes during the processes leading to the formation of the sperm or egg. Others can be stimulated by such factors as radiation, workplace hazardous chemicals, or ingested drugs, LSD for example. Whatever the cause, many of the natural abortions and non-implanting fertilized eggs are nature's way of saying that the potential child will have so many developmental abnormalities because of the chromosomal aberrancy, that it will not survive for long.

The majority of chromosomal abnormalities originate during the formation of eggs, especially in older women in the years just preceding menopause. Other significant chromosomal aberrancies occur during formation of sperm, from errors in fertilization, and from errors in the reproduction of somatic cells in the early embryo.

Importance of Formation of Gametes

Practically speaking, the process of making fully developed mature sperm from their precursor cells provides males with cells which contain genetic information and the mobility for that information to reach the egg in the upper genital tract of the female. The process leading to the formation of an egg, on the other hand, provides females once a month with a cell that also has genetic information, but, in addition contains enough nutriments and energy to carry the resultant fertilized egg through a number of cell divisions while it moves down the oviduct before it implants into the inner lining of the uterus.

This process of forming a specific cell from a more primitive cell is called *cell differentiation.* In this chapter, and in subsequent ones we will study the processes leading to the differentiation of sperm and eggs from pre-sperm cells and pre-egg cells respectively. In the chapters on the development of the embryo, we will learn about the differentiation of other types of cells in our body.

By now it should be obvious that each human sperm and egg has one set of 23 chromosomes; each set has enough genetic information to perpetuate the *genetic continuity* of the species. In addition, the processes leading to the formation of sperm or eggs afford us a way of mixing up the chromosomes of the individual producing those gametes so that the resultant sperm or eggs will have varying mixtures of chromosomes, some of which originally came from the father of the individual producing that gamete, some from the mother, and some with part of a chromosome from the mother and the other part from the father. It is this mixing up of the chromosomal make up of the sperm or egg that is a major factor leading to *genetic variability* in the offspring. The other major factor leading to genetic variability is the mixing during fertilization of the two sets of 23 chromosomes, one set from the egg and the other from the sperm, to give the full complement of 46 chromosomes in the zygote.

Basic Terms
Related to Chromosomes

Usually I dislike beginning a chapter with a list of definitions. In this case, however, as I lead you into a new subject, I think it is important to define clearly at the outset a few basic terms which we will use over and over again, and which are essential for understanding the subjects of chromosomal aberrancies, genetics, and embryological development.

Chromosomes can be defined in structural (morphological) terms as single, rodlike bodies (Fig. 9-1) found in the nucleus of a cell. In order for us to see chromosomes we need to use microscopes which have special optics while looking at a cell that is reproducing itself. We can also make the chromosomes visible for study with the microscope by applying to those cells special coloring material called "stains." Because the color-

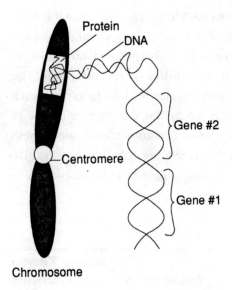

Protein

DNA

Gene #2

Centromere

Gene #1

Chromosome

Figure 9 - 1 Chromosome, DNA, and genes.

trols a specific function of a cell, such as the synthesis of an enzyme, and this capacity is passed along with the chromosome from cell to cell, and by the sperm or egg, from generation to generation.

Each gene (Fig. 9-1) can be defined in terms of biochemistry as an arrangement of certain small molecules in a chain known as *DNA*, i.e. *deoxyribonucleic acid* (Chapt. 16). Associated with the DNA in the chromosomes are a number of *proteins* serving various functions, a major one of which is to protect the DNA from harmful substances. Recall from the chapter on venereal diseases that viruses are also made up of nucleic acids and proteins. Now you can understand why scientists consider viruses to be "foreign genes" that can invade cells.

ing method has been used by scientists for quite some time, we get the name chromosome from the Greek "chromo," meaning color, and "some," meaning body.

Between both ends of the rod-shaped chromosome is a single bead-like structure called a *centromere* (Fig. 9-1). In some cases the centromere is in the center of the chromosome, but in most, it is located somewhere towards one of the ends. The centromere functions in the transport of the chromosome when the cell reproduces either in mitosis or meiosis (Chapts. 10 and 11).

Functionally, each chromosome contains a multitude of specific genes. A *gene* can be called a "unit of heredity." Each gene con-

Maternal and Paternal Origin of Chromosomes

Half of the forty-six chromosomes present in your cells are replicas of those that originally came from your mother's egg. Hence we speak of those chromosomes as being of *maternal origin*. The other twenty-three chromosomes in your cells, therefore, are of *paternal origin*. That is, they are the replicas of the same twenty-three chromosomes present in the sperm which fertilized the egg that eventually became you.

Finger Analogy

At this point, I will introduce you to a teaching tool that may help you to understand some of the cellular processes involving chromosomes. I call this tool the "finger analogy." Consider your hand a gamete and the fingers on your hand as analogous to the chromosomes present in the nucleus of that gamete. Then try to picture all of the processes and terms that we are going to describe in terms of your fingers. For example, as a start, let's say that each finger is a chromosome and that the centromere is located at the spot where the knuckle is. The rest of the finger, then, will consist of the genes and their protective proteins. I will give you some examples in a minute. Read on.

Haploid and Diploid

These terms refer to the number of sets of chromosomes a cell contains. *Somatic cells*, i.e. all cells other than the gametes, such as nerve, muscle, epithelial, etc., contain two sets of chromosomes, one set of maternal origin and another of paternal origin, and are called *diploid*. A fully differentiated gamete, however, has only one complete set of chromosomes and, therefore, is called *haploid*. A fertilized egg (zygote) has one set of chromosomes from the egg and a second set from the sperm, and is therefore, diploid.

The number (n) of chromosomes in the haploid nucleus of a gamete varies from species to species. Humans have twenty-three chromosomes per gamete. Thus, in humans, n = 23, and a human diploid cell would have 2n or 46 chromosomes (2 X 23). Other animals usually have different n numbers, that is a different fixed number of chromosomes in their gametes.

Finger Analogy for Haploid and Diploid

To help visualize a haploid and a diploid cell via the finger analogy, assume the n number for your "hand gamete" to be 5. Thus, the fingers on your right hand could represent the five chromosomes in a haploid sperm, and the fingers on your left hand the five chromosomes in a haploid egg. To illustrate the diploid number, simply clasp the fingers of your right and left hands together and now you have the 2n number of chromosomes (diploid), or 2 X 5 = 10; five of those chromosomes by definition are of paternal origin, and the other five of maternal origin.

Homologous Pairs of Chromosomes

Diploid cells have two sets of chromosomes, each set being identical to the set originally donated by the gametes from each parent. Thus, for example, each diploid cell containing a long chromosome that originally came from the 23 chromosomes present in the mother's egg, will also contain a long chromosome of the identical genes that came from the father's sperm; we call those two long chromosomes homologous chromosomes. The same applies to every chromosome in the set of 23 donated by the mother; for each one, there will be a *homologous* chromosome of identical structure and gene composition in the set of 23 chromosomes contributed by the father.

Figure 9-2 shows this situation for six chromosomes; the chromosomes of maternal origin are black whereas those of paternal origin are white. Thus, a human somatic cell has 23 pairs of homologous chromosomes.

Figure 9 - 2 Diploid cell (n=3; therefore 2n=6)with three chromosomes of maternal origin (black) and three of paternal origin (white).

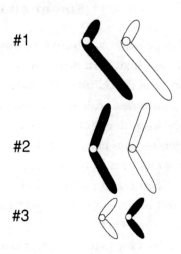

Figure 9 - 3 Three pairs of homologous chromosomes from cell depicted in Figure 9 - 2.

Each pair of homologous chromosomes is designated by a number ranging from 1 to 23, one for each pair. Figure 9-3 shows three pairs of homologous chromosomes taken from the cell drawn in Figure 9-2).

Finger Analogy for Homologous Pairs of Chromosomes

To help visualize homologous pairs of chromosomes via the finger analogy, assume that instead of 46 chromosomes, the diploid somatic cell has 10 chromosomes, five different ones from the father, and five homologous ones from the mother. That is, using the finger analogy, the hand representing a father's sperm would have a thumb (T_f), forefinger (F_f), middle finger (M_f), ring finger (R_f) and little finger (L_f) chromosome. The subscript f stands for father. Likewise, the hand representing a mother's egg would contain one T_m, one F_m, one M_m, one R_m and one L_m chromosome, the subscript m standing for mother. Thus, in the diploid cell represented by a hand clasp, the homologous pairs of chromosomes are: $T_f\, T_m$, $F_f\, F_m$, $M_f\, M_m$, $R_f\, R_m$, and $L_f\, L_m$. In the nucleus of a real cell, the chromosomes are mixed up in a random fashion.

Alleles

Before we get deeper into terms dealing with chromosome structure, we need to remember that we study chromosomes primarily because of their genetic capacities, i.e. because of the genes they contain. Recall that a cell needs only one of each kind of gene to survive. Yet diploid cells have two sets of chromosomes, and, therefore, a double dose of each gene. For each gene on a chromosome originally provided by the mother, there is a comparable gene at the homologous chromosome provided by the father. These two genes which are at the *same location* on each of the chromosomes of a pair of homologous chromosomes, and which *control for the same function* are called *alleles*, or an allele set. Can you think of why we are

better off by having alleles, i.e. a pair of genes to control a function of the cell, rather than a single gene? I will give you the answer later in this chapter.

Finger Analogy of Alleles

Using the finger analogy again, let us say that fingers F_m and F_f (i.e. your two forefingers) represent a pair of homologous chromosomes in a cell of an adult. With a felt pen, make a point on your right forefinger (for example, in the middle of the fingernail) and we will say that point represents the gene for sickle cell anemia on one chromosome (F_m). Thus, the allele would be the gene located at the same spot on the fingernail of the left forefinger (F_f). Therefore, make a point of the same color and size there with the felt pen. In the finger analogy, we shall consider the two genes represented by the two marked dots to be alleles, or an allele set of genes; i.e. they are on the same location on homologous chromosomes and control the same function.

Mutation

What happens if one of the alleles, i.e. one of those two genes, changes? We call a permanent chemical change in a gene a *mutation*. That is, the order of the molecules in the

DNA making up that gene becomes altered in such a way that the gene has changed its original function and, in many cases, has no function at all. A mutated gene is carried on its chromosome from generation to generation. When both genes of an allele set mutate to become nonfunctional, the person having those mutated genes is usually affected in some harmful way and suffers from a *genetic disease* (see Chapters 14 and 16). In other cases, the mutation may lead to a new function that may prove advantageous to the species.

Mutation Visualized by the Finger Analogy

Refer to the dark points representing the two genes of an allele set that you marked on the nail of each of your forefingers. Let us say that if the marks indicating those genes were made with a black felt pen, then the genes are normal. On the other hand, if one of those genes mutated (now change the black mark on one of the fingers to a red mark), then the "red" gene would be different from the normal (black) one. Both genes (the black and red ones) are still genes of the same allele set, that is they are still on the same spot on the two chromosomes, and still affect the same function. Nonetheless, one of those genes (indicated by black) functions normally, and the other (indicated by red) has mutated.

The mutated gene, however, may affect that function in an abnormal way, if the function is retained at all. To use the finger analogy to represent a genetic disease, you would mark both genes of the allele set in red, indicating that neither gene would function normally.

Autosomal and Sex Chromosomes

Only one chromosome in a set of 23 (or 2 in a diploid cell having 46 chromosomes) is involved in determining the sex of the individual. In humans, the so called *sex chromosome* is designated by either an X, if it controls female traits, or a Y if it controls for traits of maleness. In human cells, the X chromosome is much larger than the Y chromosome. Individuals whose cells have two X chromosomes develop as females whereas those with an X and a Y chromosome develop as males. All the other chromosomes in the cell do not deal with controlling the sex of the individual; these are called *autosomal* (non-sex) *chromosomes*.

Chromatids in Doubling-Chromosomes

A *chromatid* refers to each of the two strands of a chromosome while it is duplicating itself. Recall that a single chromosome is a rodlike structure having one centromere located somewhere between the ends (Fig. 9-1). Chromatids, on the other hand, exist in a chromosome that is doubling. Each chromatid of a doubling-chromosome shares a single centromere (Fig. 9-4). Put another way, a *doubling-chromosome*, which is composed of two chromatids held together by one centromere, is in reality one chromosome that has doubled the rod part of its structure prior to the reproduction of the cell that takes place in mitosis or meiosis (Chapters 10 and 11). Once the doubling-chromosome stage with its two chromatids is over, each doubling-chromosome reproduces its centromere and starts separating immediately before cell reproduction into two complete and identical chromosomes, each consisting of one rod and one centromere.

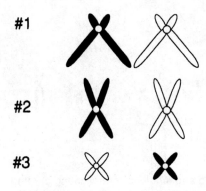

Figure 9 - 4 Three pairs of homologous doubling - chromosomes from cell depicted in Fig.9 - 2 while it is reproducing.

Karyotype

A *karyotype* is a pictorial record of photographs of the chromosomes found in one somatic cell; the picture is structured so that the photographs of the chromosomes are arranged in pairs of homologous chromosomes according to their length and the location of their centromere. Figure 9-5A shows the 46 chromosomes of a human male. One chromosome of each pair came originally from the father of the person whose cell was analyzed, and the other from that person's mother. But because the homologous chromosomes in a pair look alike, we can not tell the maternal or paternal origin of each autosomal chromosome simply by looking at it. In the case of males, however, we can be certain that any Y chromosome we see comes from that person's father and that the X chromosome originally was contributed by the mother.

Now that you think you understand karyotypes, let me tell you the trick I played. The karyotype drawn in Fig. 9-5A is not really true to life because it is not possible to see and photograph only the single chromosomes in a cell. As you will learn in Chapters 10 and 11, the only time that chromosomes of a cell are visible microscopically is when they are doubling, i.e. when the chromosomes are in the doubling-chromosome stage. Thus, the true karyotype that you will see consists of forty-six doubling-chromo-

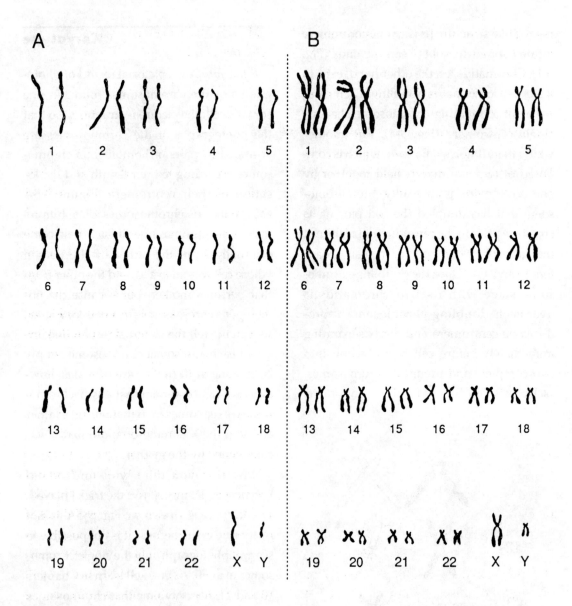

Figure 9 - 5 A. Theoretical "Karyotype" if it were possible to see single non-duplicating chromosomes. B. The 23 pairs of doubling-chromosomes of a normal male. Twenty-three of the doubling-chromosomes are of paternal origin, and 23 are of maternal origin, although we cannot tell which merely from looking at them. All that we know for sure is that the Y chromosome comes from the father. We call such a display of a person's doubling-chromosomes a karyotype. Once during mitosis or meiosis when each centromere duplicates and each of the chromatids of a doubling-chromosome separate, the separated "chromatids" become bonafide chromosomes.

somes. While in the doubling-chromosome stage, each chromosome has sort of an "X" appearance (Fig. 9-4). Examine Figures 9-5A and 9-5B; note that in both figures, the chromosomes (Fig. 9-5A) or doubling-chromosomes (Fig. 9-5B) are still arranged by their size and location of their centromere. The "X-like structures" shown in the true karyotype (Fig. 9-5B) show the two chromatids of a single doubling-chromosome linked at the commonly shared centromere.

We can learn much from examining the karyotype of a cell. For one, as indicated in the foregoing, we can tell the sex of the individual from karyotypic analysis. Note that in the lower right hand corner of Figure 9-5B there is one doubling X chromosome and one doubling Y chromosome, indicating that the cell came from a male. If the karyotype showed two doubling X chromosomes, then the cell would have come from a female.

Identification of the sex of the donor may be important when diagnosing cells taken from the amniotic fluids of a prospective mother by a process called amniocentesis (see Chapt. 12). Similar examinations of the karyotype of cells taken from amniotic fluids also may help diagnose, for example, cases of Down's syndrome (an extra number 21 chromosome) or Turner's syndrome (absence of one X chromosome). Such diagnoses can give the parents crucial information for deciding on abortion or on having a child with a particular chromosomal abnormality

and with all of the characteristics that go with that abnormality.

Why Two Sets of Chromosomes?

Although this chapter appears to be concerned mostly with definitions, it also hints at the heart of human genetics - the reason why we have diploid somatic cells containing twice the necessary number of chromosomes and genes. The answer becomes clear when we consider what would happen if our somatic cells were haploid, not diploid, and one or more genes mutated so that they became non-functional. In such a case, we would invariably suffer a genetic deficiency in which a gene would not function properly and, as a consequence, an important protein, such as an enzyme, would be lacking.

Our cells are diploid, however, and we do not suffer a genetic deficiency if only one of the genes of the set making up an allele mutates. Why? Consider the possibilities. If only one gene of the allele set mutated, that gene would not function; but the other gene in the allele set which has not mutated will function normally. In most cases, if only one gene of an allele set works, that gene will carry out the function of the cell irrespective of the non-functioning mutated gene. Hence, by having a pair of genes (an allele set), by being diploid, by having homologous pairs of chromosomes, we have an extra measure of insurance to guarantee that our bodies and that the bodies of our children will still function properly even should some mutations arise.

On the other hand, if both genes of allele set mutate, then the person harboring those mutated genes usually will suffer from a genetic disease. How does a person usually inherit a genetic deficiency? Genetic deficiencies occur when each of the parents, who may appear normal in all outward respects, carries one mutated gene of the same allele set, and when the haploid sperm and the haploid egg from those parents both carry the mutated gene as they unite to form a fertilized egg. More about this subject when we cover genetics (Chapts. 13-15) and genetic counseling (Chapt. 15).

A Brief Introduction to Molecular Biology

I have just introduced you to key terms involving the structure of chromosomes. Now I will give you a brief introduction to some terms and concepts we commonly use in describing the molecular biology of gene action. Chapter 16 will cover the same material in more detail.

The essential fact in the molecular biology of genetics is that the gene, which is made of the nucleic acid called *DNA*, deoxyribonucleic acid, has the information to make a specific protein with the help of another kind of nucleic acid called *RNA*. If the gene makes a functional protein, we call that gene a *dominant* gene. If the gene makes a protein that is not functional, we call that gene a *recessive* gene. In outline form, we would write:

Gene ⟶ Via RNA ⟶ Protein
(DNA)   (e.g. an enzyme)

GENE Dominant ⟶ Via RNA ⟶ Protein
(DNA)   (Functional)

GENE Recessive ⟶ Via RNA ⟶ Protein
(DNA)   (Nonfunctional)

So far so good? What? You do not know what a protein and an enzyme are? Very simply, a *protein* is a chain of amino acids which are linked together and which fold and coil in a specific pattern dependent upon the sequence of the amino acids in that protein. An *enzyme* is a protein that acts as a catalyst to speed up the rate of a specific chemical reaction taking place in our bodies.

Now you may begin to see the "chicken and the egg" problem of writing genetics textbooks. How can we understand what a protein is if we do not know what an amino acid is?

Amino Acids

An *amino acid* can be defined as a small molecule having a chain of carbon atoms, usually from one to five, in the middle, with an amino group at one end and an acid group at the other. Figure 9-6A gives two simple ways to illustrate the basic structure of an amino acid. At the top of the figure you will see a box labeled carbon chain. At the right end is the letter N to symbolize the *amino group* because its main component is the nitrogen atom. At the other end is the letter H to symbolize the *acid group* because its main component is hydrogen. The clothes-pin-like drawing at the bottom of the figure is an even more abstracted simplified diagram of the basic structure of an amino acid, with the peg end representing the amino group and the forked end representing the acid group.

There are over twenty amino acids necessary for life. Each of them has the same basic structure, as indicated in Figure 9-6A, and in addition, has a side group attached to the center of the carbon chain that is in between the amino and acid groups. The side groups (Fig. 9-6B) may have a positive charge (+), a negative charge (-), a neutral charge (o), or may consist of sulfur groups. A mixture of these amino acids links up in a specific order (Fig. 9-6C) directed by the gene (DNA) with the help of RNA. Finally, each chain of amino acids folds and coils up in a specific manner because of the attractions of the various side groups on the amino acids making up that protein (Fig. 9-6D).

Proteins

Examine Figure 9-6D carefully and you will note that the negative groups are attracted to the positive groups, the neutral groups, like two oil droplets, are attracted to each other, and the sulfur groups link up to each other. It is this specific shape of the folded and coiled protein that gives each specific protein its function. For example, the protein which becomes an enzyme that digests carbohydrates has a specific order of amino acids and a specific shape. Likewise, the protein which becomes part of the molecule hemoglobin, needed to transport oxygen in our red blood cells, has a specific order of amino acids and a specific shape. As you will learn in Chapter 16, if only one amino acid in the chain making up the hemoglobin protein is altered, we get a condition known as sickle cell anemia.

Example of Enzyme Action

Most of the proteins made by cells are *enzymes*, the catalysts that speed up the rates of chemical reactions. Without enzymes, the cells would not function. For our example,

A

Two ways to illustrate an amino acid.

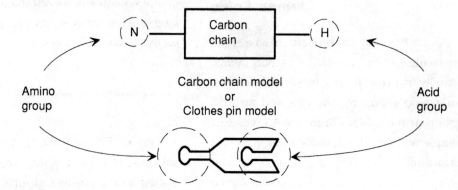

Amino group

Carbon chain model
or
Clothes pin model

Acid group

B

Each amino acid has a different side group on it.

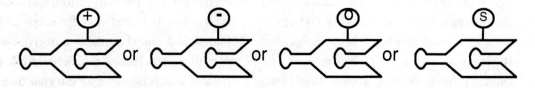

C

They link up (via DNA + RNA) to form a chain of amino acids.

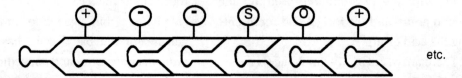

etc.

D

Because of side groups, the chain of amino acids coils up to make a protein.

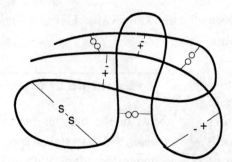

Figure 9 - 6

let us look into the ways enzymes in the follicle cells participate in the synthesis of estrogen.

The steroid hormone estrogen is a complex molecule (Fig. 9-7). The synthesis of estrogen requires the involvement of eight different enzymes, indicated by E_A through E_H. First enzyme A catalyzes the combining of two molecules of acetic acid (vinegar), each molecule containing two carbon atoms, to produce a molecule consisting of four carbon atoms. Then, through a series of reactions catalyzed by enzymes B through F, a number of those four carbon molecules condense to manufacture a carbon structure shaped like four joined polygons. Enzyme G

converts that structure to testosterone, and enzyme H converts the testosterone into estrogen and vice versa.

Now you see why females can secrete some testosterone and why males can secrete some estrogen. Both the follicle cells and the interstitial cells possess enzymes A through G. The follicles, on the other hand produce a good deal of enzyme H whereas the interstitial cells produce only small amounts of this enzyme. In addition, for every enzyme participating in the synthesis of testosterone and estrogen, there is a specific gene controlling the synthesis of that enzyme. Thus, if the gene which controls the synthesis of enzyme H in the follicle mu-

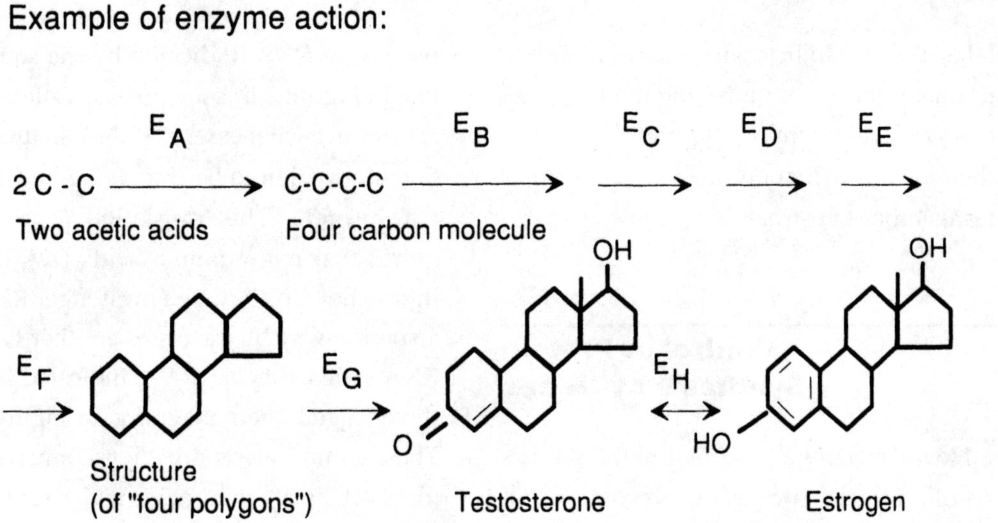

Example of enzyme action:

Figure 9 - 7 Schematic drawing of enzymatic synthesis of testosterone and estrogen from the two carbon molecules of acetic acid.

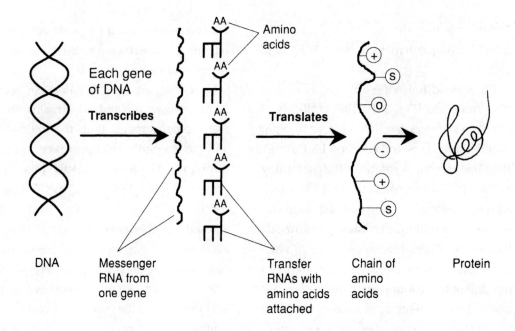

Each gene
of DNA

Transcribes

Amino
acids

Translates

| DNA | Messenger RNA from one gene | Transfer RNAs with amino acids attached | Chain of amino acids | Protein |

Figure 9 - 8 Schematic summary of the synthesis of a protein controlled by a gene (DNA).

tates, then the follicle will make only testosterone! If the gene controlling the synthesis of any of the enzymes A through G mutates, then the cell will not be able to make either testosterone or estrogen.

Control of Protein Synthesis by Genes

How then does a gene control the synthesis of a protein, such as an enzyme? Recall that the gene is a specific arrangement in the long DNA molecule (Fig. 9-1). Each gene in the DNA can form a molecule that we call

messenger RNA (indicated by the squiggly line in Figure 9-8) by a process called *transcription*. Each messenger RNA, in turn, has the information to be *translated* into a specific protein. The translation process requires that many amino acids (AA in the figure) line up along the messenger RNA in a specific way by means of another type of RNA called *transfer RNA*, indicated by the three-legged chair diagrams in Figure 9-8. Those amino acids are then connected to form a chain of amino acids which fold up to form a protein as in Figure 9-6D. Figure 9-8 shows one such linked chain of amino acids folding up to form a specific protein originating from a specific gene and messenger RNA. If something goes wrong in the synthesis of the protein to give a protein differing in only one amino acid, then that new protein would be considered as the product of a mutation.

All of these concepts are explained more fully in Chapter 16. Nonetheless, with this general introduction, you are now prepared to study genetics and the formation of gametes.

Chapter 10

Reproduction of Cells: Mitosis and Meiosis

Overview

Mitosis and meiosis refer to two different processes by which the cells in our bodies increase in number. In *mitosis*, our *somatic* cells, which are *diploid*, reproduce to form more diploid somatic cells; these cells, the original ones and the new ones, are genetically the same. Mitosis also takes place in the initial divisions of the *fertilized egg*, and in the *pre-egg cells (oogonia)* and *pre-sperm cells (spermatagonia)*.

Meiosis, on the other hand, does not refer to somatic cells, but only to the final two divisions of the pre-egg and pre-sperm cells that develop into the haploid *gametes*, i.e.

either eggs or sperm. Thus, in meiosis, we go from *diploid (2n)* cells to *haploid (1n)* ones. Both of the two haploid cells formed from a meiotic division have 23 chromosomes, but those cells are not genetically the same as is the case with the somatic cells originating by mitosis.

For either mitosis or meiosis to take place, the chromosomes must double, and therefore go through the doubling-chromosome stage. We can simplify the fate of the chromosomes in mitosis as:

**Somatic cell with
46 chromosomes
(2n)**

↓

**Same cell with
46 doubling-chromosomes
(2n)**

**(Daughter cell)
46 chromosomes
(2n)** **(Daughter cell)
46 chromosomes
(2n)**

The fate of the chromosomes in meiosis is somewhat more complicated and goes through two cell divisions to get to the final haploid gametes:

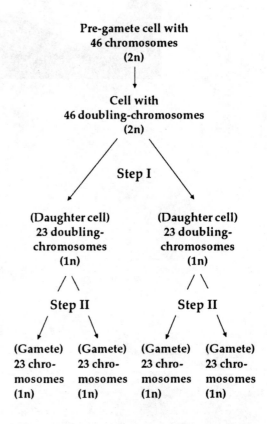

**Pre-gamete cell with
46 chromosomes
(2n)**

↓

**Cell with
46 doubling-chromosomes
(2n)**

Step I

**(Daughter cell)
23 doubling-
chromosomes
(1n)** **(Daughter cell)
23 doubling-
chromosomes
(1n)**

Step II **Step II**

**(Gamete)
23 chro-
mosomes
(1n)** **(Gamete)
23 chro-
mosomes
(1n)** **(Gamete)
23 chro-
mosomes
(1n)** **(Gamete)
23 chro-
mosomes
(1n)**

Although meiosis is more complex than mitosis, many of the stages in meiosis, but not all, are similar to those taking place in mitosis. Thus, I will first introduce you to the simpler and basic process of mitosis. This information should prepare you for a clearer understanding of meiosis and egg and sperm formation.

Mitosis

Because mitosis leads to the formation of many cells, each having the same diploid number of chromosomes (2n) as does the parent cell, mitosis allows for *genetic uniformity* in the somatic cells of an organism. In other words, as a result of mitosis, all of the new cells have the identical chromosomes and genes as were found in the parent cell. Exceptions do occur, of course, as when one or more of the genes mutate just prior or during mitosis. Other exceptions are explained in Chapter 12 .

Mitosis also takes place each time the fertilized egg becomes the 2-cell, 4-cell, 8-cell, 16-cell etc. until about the 128 cell stage (Chapt. 17). After each of these mitotic divisions, the cells becomes smaller and smaller. The number of chromosomes in the nucleus of each cell, however, stays the same, i.e. at the diploid number of 46. In contrast, the yolk-rich cytoplasm outside of the nucleus in these cells becomes smaller and smaller until a relatively constant ratio of *nuclear volume to cytoplasmic volume* is reached. Hence, because each successive cell becomes smaller and smaller, we speak of the mitoses of the fertilized egg as *cell divisions,* although the nucleuses and chromosomes multiply each time two new cells form.

Once the 128 cell stage of the dividing fertilized egg (called the *blastocyst,* Chapt. 18) implants into the wall of the uterus, and gets food there, the subsequent mitotic processes taking place in the somatic cells of the implanted blastocyst lead to more diploid cells of the same size, and the embryo starts to enlarge with each successive increase in the number of cells. Hence, for somatic cells in general, mitosis should more properly be called either *cell reproduction* or *cell multiplication.*

Cellular Structures Involved in Mitosis

The major structures involved in the reproduction of cells can be seen in a simplified diagram of a typical somatic animal cell (Fig. 10-1). The *chromosomes* with their *centromere* are located inside the *nucleus.* The

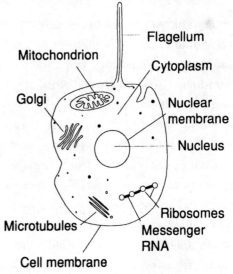

Figure 10 - 1 Components of typical somatic animal cell.

nucleus is bounded by a *nuclear membrane* that breaks down during mitosis and is reconstructed as two new cells form. The rest of the cell outside of the nucleus is called the *cytoplasm*. The cytoplasm is also bounded by a membrane called the *cell membrane*. In the cytoplasm are many components that aid in cell multiplication, some giving energy and others raw materials. Also in the cytoplasm are a number of tubes of extremely small diameter *(microtubules)* which have a number of functions including that of becoming part of the *spindle* (see below) in mitosis and meiosis.

Mitosis

When studying mitosis, most of our attention should focus on the number and position of the chromosomes while the cell is reproducing to form two daughter cells. Some call mitosis "nature's choreography" or "the dance of the chromosomes," because the movement of the chromosomes during mitosis is precisely arranged and is a magnificent sight to behold, especially when seen magnified in movies played at high speeds. It takes approximately 12 hours for a human somatic cell to reproduce once mitotically.

Biologists arbitrarily divide mitosis into six stages, and each is given a scientific name. In this book I label the stages by number and with an English word or phrase. For those of you who wish to read more about mitosis in other books I give in parentheses the scientific names of each stage. In the illustration (Fig. 10-2), to simplify matters, only four of the chromosomes are shown rather that the forty-six normally present in human somatic cells. Hence, we can say that the diploid cells in this illustration are 2n with n being 2 rather than 23 as is the case with human somatic cells.

Stage I: Preparatory Stage (Interphase)

In this stage the cell prepares to divide by manufacturing more DNA, thereby enabling the chromosomes to duplicate themselves by forming doubling-chromosomes. Because these preparations involve chemical steps, and, therefore, can not be seen with the microscope, biologists formerly gave this stage little importance and called it interphase. Actually we now know that this is a most important stage. Because DNA is made in this stage, it is the one in which radiation and carcinogens, for example, have their greatest harmful effects by altering the structure of the DNA while it is being manufactured.

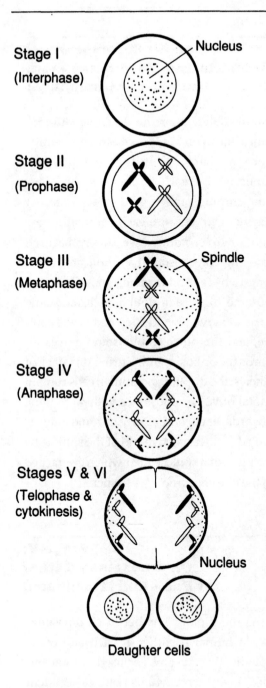

Stage I
(Interphase)

Nucleus

Stage II
(Prophase)

Stage III
(Metaphase)

Spindle

Stage IV
(Anaphase)

Stages V & VI
(Telophase &
cytokinesis)

Nucleus

Daughter cells

Figure 10 - 2 Stages of mitosis of a cell having four chromosomes (i.e. n = 2, or 2n = 4).

Stage II: Appearance of Chromosomes (Prophase)

During this stage the chromosomes thicken and become visible when viewed either by means of colored dyes applied to the cell, or by means of a microscope outfitted with special optics. Not only are they visible, but each already appears as a "four-armed" doubling-chromosome linked by a single centromere. The membrane around the nucleus breaks down during this stage so that you can no longer see it, and the microtubules, now called *spindle fibers*, start to form the spindle. The *spindle* is a web-like structure shaped like a football whose outer shell is made up of a network of *microtubules* which go from one end of the "football" to the other (Fig. 10-2). Each of the 46 doubling-chromosomes attaches by its centromere to a single spindle fiber of the spindle. When all of the doubling-chromosomes are attached to their own spindle fiber, the doubling-chromosomes start moving towards the center ("equator") of the spindle and away from the two ends (poles).

Stage III: Lining up of the Chromosomes at the Center (Metaphase)

During this stage all of the doubling-chromosomes are lined up along the equator of the football-shaped spindle. The forty six doubling-chromosomes are arranged one by one on separate spindle fibers so that they can be found all along the circumference of the equator.

There is no particular order of the doubling-chromosomes at the equator: (a) the doubling-chromosome of chromosome 1, for example, is not necessarily next to the doubling-chromosome of chromosome 2; (b) the 23 doubling-chromosomes of maternal origin are not lined up separate from the 23 doubling-chromosomes of paternal origin. (c) the homologous doubling-chromosomes of maternal origin do not line up next to the homologous doubling-chromosome of paternal origin. In other words, during Stage III of mitosis, the doubling-chromosomes line up at the equator of the spindle *randomly* with one doubling-chromosome per spindle fiber. You will find out later (Chapt. 11) that this type of arrangement, that is, one doubling-chromosome per spindle fiber, does not occur during the first of the two meiotic divisions.

Scientists have found that by adding a certain drug to cells that are multiplying, they stop mitosis from proceeding past Stage III. They use cells stopped during this stage to prepare karyotypes of the chromosomes. Because chromosomes always appear as doubling-chromosomes in Stage III, the karyotypes, therefore, also show chromosomes as doubling-chromosomes (Fig. 9-5B).

Stage IV: Chromosomes Separate and Move Away From Each Other (Anaphase)

Immediately after the doubling-chromosomes line up along the equator, the centromere of each doubling-chromosome duplicates and each chromosome of a doubling-chromosome begins to separate. At this moment, because each chromatid now has its own centromere, we can say that each doubling-chromosome has duplicated to become two identical chromosomes. One chromosome from each doubling-chromosome starts to move towards one end of the spindle, and the other chromosome moves towards the opposite end. During this stage of mitosis, the duplicating human cells contain a total of 92 chromosomes with 46 moving towards each end. Each chromosome is pulled to its respective end of the spindle by the piece of a spindle fiber which is attached to the chromosome's centromere.

Stage V: Chromosomes Gather at Each End (Telophase)

As the diploid number of chromosomes (46 in humans) gather at each end of the spindle, they become enclosed in a new nuclear membrane. At this point the cell temporarily has two nucleuses, each having a diploid number of chromosomes.

Stage VI: Partitioning of the Cell into Two Cells (Cytokinesis)

The membrane surrounding the cell forms a furrow around the circumference of the cell just above the place where the equator of the spindle was located during Stage III. This furrow finally constricts so that the original cell separates into two daughter cells, each having identical diploid sets of chromosomes. Thus, except in the case of a mutation or of a chromosomal aberrancy, these two cells formed by mitosis will have the identical chromosomes and genes.

Significance of Cell Reproduction by Mitosis

Because the two daughter cells have identical chromosomes and genes, we can see that the human being derived from the many mitoses starting from a single fertilized egg will have cells that are *genetically uniform*. Such uniformity leads to a normal and orderly development.

If the cells derived from mitosis have the same chromosomes and genes, how can we account for the diversity of cell types found in an individual? Why are not all the cells the same? The answer to these questions lies in the fact that different genes in different cells are turned on or off at different times during the development of an individual (Chapt. 16). For example, during the formation of the embryo (Chapt. 20) the genes in some cells are directed to become nerve cells, whereas the genes in other cells are directed to become liver cells. Such is the case for all differentiated cells in the body.

What Can Go Wrong During Mitosis?

The reproduction of cells in an individual does not always continue in the coordinated sequence described under mitosis. A number of things can go wrong, some of which we understand, some of which we do not. In the latter category there is cancer. In the simplest of terms, in a cancerous growth, some cells which normally have stopped reproducing start to reproduce in an uncontrolled fashion. Such uncontrolled cell reproduction may lead to the formation of undifferentiated growths that often form obstructions in the body, blocking key blood vessels and nerves, leading to pain and finally to death. Although much progress has been made in controlling cancers, we do not always know how to control the reproduction of cancer cells. Usually any agent we use to stop cancer cells from reproducing, such as radiation and certain anti-cancer drugs, will also stop other cells in the body from reproducing. That is one reason why cancer is such a difficult illness to control.

Mosaics

Other anomalies of mitosis can affect us. One of these is the formation of cells with an abnormal number of chromosomes. When such cells multiply many times in a person to form a group of these atypical cells, we say that the person exhibits a *mosaic* condition. Thus, we can define a mosaic as an organism having groups of cells in various parts of its body which have a different chromosomal or genetic make-up than do the cells adjacent to those groups. There are various ways that chromosomal mosaics can develop.

As an example, I discuss here one such case, a mosaic forming because of an accident of mitosis called *chromosome lag* (or *anaphase lag*). In chromosome lag (Fig. 10-3), one chromosome lags behind the other chromosomes when one group of 46 is moving towards the end of the spindle during Stage IV, the separation phase, of mitosis. If such a chromosome lags too far behind while the other chromosomes migrate toward the end of the spindle, it may be left behind in the cytoplasm as the new nuclear membrane forms around the rest of the 45

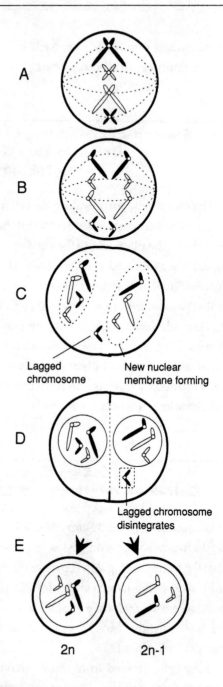

Lagged chromosome

New nuclear membrane forming

Lagged chromosome disintegrates

2n   2n-1

Figure 10 - 3 Chromosome (Anaphase) lag during mitosis.

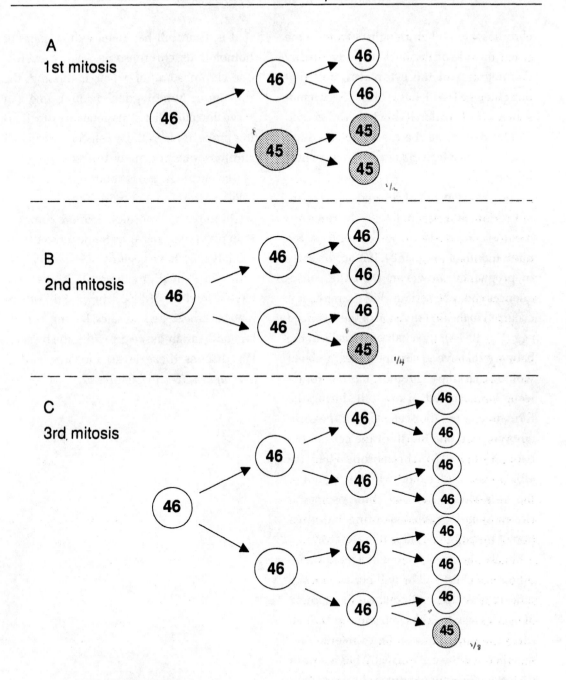

Figure 10 - 4 Examples of mosaics forming if chromosome lag takes place during: A, first mitosis; B, second mitosis; and C, third mitosis. The numbers represent number of chromosomes in the nuclei of these somatic cells.

chromosomes. When a lagging chromosome is left outside of the nucleus, it eventually disintegrates in the cytoplasm, and that daughter cell will have only 45 chromosomes (2n - 1) within its nucleus, rather than 46 (2n) as will be the case with the other daughter cell which did not experience chromosome lag.

The consequences of chromosome lag are not serious if it occurs late in the development of an individual, for example in the ninth month of pregnancy. If it occurs early in pregnancy, however, then the consequences could be serious. For example, if it occurred in the first division of the fertilized egg (Fig. 10-4A), then half of the cells in the baby would have 46 chromosomes, and half would be lacking a chromosome. If chromosome lag occurred in one cell during the formation of the four-cell stage of the early embryo, then one fourth of the cells in the baby would have 45 chromosomes (Fig. 10-4B). Do you understand why one eighth of the cells would have 45 chromosomes if chromosome lag occurred during the formation of the eight-cell stage (Fig. 10-4C)?

Finally, the development of a mosaic condition in a child will or will not have a significant effect depending upon the number of its cells lacking a chromosome and which chromosome is missing. Of course you remember that the cell normally has a pair of each of the 23 kinds of human chromosomes, one from the mother and one from the father;

that is, each cell has twenty-three pairs of homologous chromosomes. Thus, even if one chromosome of a pair is missing, the remaining homologous chromosome can serve the cell well if all its genes are functioning properly. If in those cells possessing 45 chromosomes, a gene of the remaining unpaired chromosome is abnormal, then those cells may not function normally.

Chromosomal mosaics are more common than previously thought. Some mosaic individuals may have regions of the body that differ in skin pigmentation. Others may have a mixture of blood types. Still others may become sexual mosaics, having testicular and ovarian tissue growing side by side. We discuss these latter mosaics, called *hermaphrodites*, in Chapter 12.

Chapter 11

Meiosis

We have already learned that mitosis is a method whereby a somatic cell reproduces two daughter cells, each having an identical diploid set of chromosomes. Mitosis imparts *genetic uniformity* to the cells of an organism, i.e. all the cells get the same kind and number of genes in their chromosomes.

Meiosis, on the other hand, refers to the process whereby a single diploid cell by means of two cell divisions gives rise to four *haploid* daughter cells. Because the chromosomes in the resultant cells are reduced in number from forty-six to twenty-three after the two meiotic divisions (i.e. the cells go from the diploid to the haploid state), meiosis is often called *reduction division.* Furthermore, meiosis occurs only in the final two cell divisions leading to the formation of gametes. Meiosis does not occur in somatic cells.

Meiosis, in contrast to mitosis, provides both for genetic continuity of a species from generation to generation, and for genetic variability among the progeny. By *genetic continuity*, we mean that the same basic set of haploid chromosomes with their respective set of genes are passed on from generation to generation by means of the uniting haploid sperm and haploid egg.

Meiosis also provides for *genetic variability* because the progeny will get from each parent differing proportions of the parent's maternal and paternal chromosomes (i.e. differing proportions of the genes originally contributed to the two parents by the four grandparents). Hence, as a result, no two individuals, unless they are identical twins, will get exactly the same chromosomes and genes. This variation in the genetic makeup of all individuals created through sexual reproduction, allows for evolution, and, therefore, for the survival of the species.

An Overview of Meiosis

Meiosis consists of two successive cell divisions of a diploid pre-egg or pre-sperm cell which give rise to four haploid daughter cells (Fig. 11-1). Before the first division begins, however, the forty-six chromosomes of the diploid cell duplicate to form forty-six doubling-chromosomes, just as in mitosis. As a result of the first division, therefore, each daughter cell gets a haploid number (twenty-three) of doubling-chromosomes. In the second meiotic division, the twenty-

three doubling-chromosomes separate into two sets of twenty-three chromosomes, with just one set, i.e. the haploid number of single chromosomes, going to each of the daughter cells.

First Meiotic Division (Meiosis I)

The preceding paragraph states that the diploid cell first forms a diploid set of forty-six doubling-chromosomes. In other words, in Stage I of Meiosis I, the cell duplicates its DNA, i.e. its genes. This process always occurs prior to cell division whether it be by meiosis or mitosis. In Stage III of the first meiotic division, however, the forty-six doubling-chromosomes do not line up at the equator one per spindle fiber as they do in mitosis (Fig. 11-2A). Instead, in Stage II of the first meiotic division each doubling-chromosome "finds" its homologous doubling-chromosome, and pairs up with it on the same spindle fiber at the equator of the spindle (Fig. 11-2B). In addition, unlike the process in mitosis, the centromere of each doubling-chromosome does not reproduce itself in Meiosis I. To the contrary, instead of forty-six single chromosomes going to each end of the spindle in Stages IV and V, as they do in mitosis, in Meiosis I twenty-three doubling-chromosomes go to each end. The products of the first meiotic division (Fig. 11-1) of

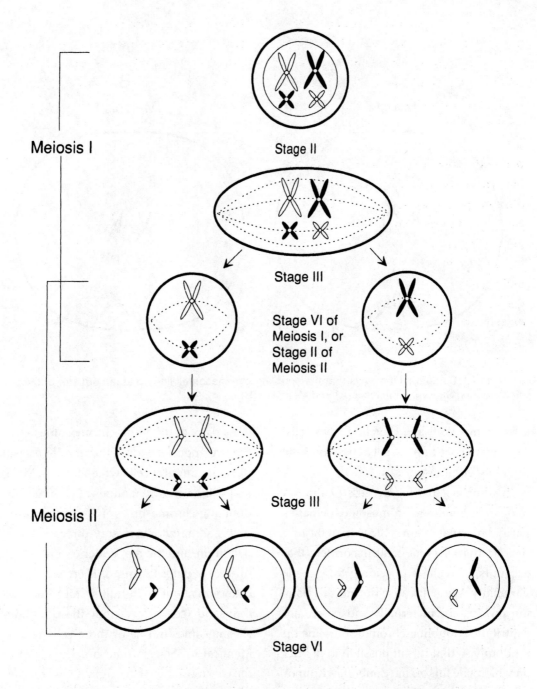

Meiosis I

Stage II

Stage III

Stage VI of
Meiosis I, or
Stage II of
Meiosis II

Stage III

Meiosis II

Stage VI

Figure 11 - 1 The two meiotic divisions of a diploid pre-gamete cell. In this example n = 2 and 2n = 4. The black chromosomes represent those of maternal origin, and white, of paternal origin.

A
Mitosis

B
Meiosis I

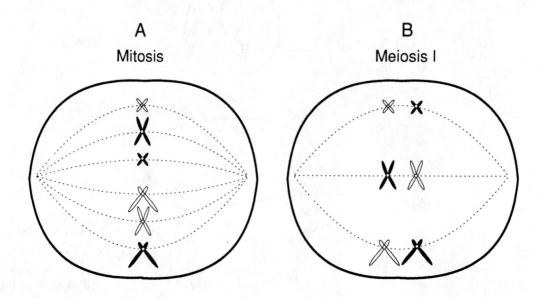

Figure 11 - 2 Comparing the ways that the doubling-chromosomes line up at the equator of the spindle during Stage III of mitosis (A) and Meiosis I (B).

a pre-gamete cell are, therefore, two cells, each having one haploid set of twenty-three doubling-chromosomes.

This unusual pairing process of the homologous doubling-chromosomes of maternal and paternal origin in Stage III of the first meiotic division is one of the major steps that accounts for meiosis providing for genetic variability. Why? Because during this pairing process, the maternally and paternally contributed doubling-chromosomes line up randomly so that it is impossible to predict on which side this or that doubling-chromosome will end up. Thus, when the homolo-

gous pairs of doubling-chromosomes go to the two opposite ends of the spindle, each of the two daughter cells which form in Meiosis I will have combinations of twenty-three doubling-chromosomes of diverse mixtures, that is, some (zero to twenty-three) of maternal origin, and the remainder (twenty-three to zero) of paternal origin. There is only one chance in about eight million four hundred thousand (see below) that the doubling-chromosomes in one of those cells will be identical in their chromosomal makeup to the chromosomes that originally were in either the egg or sperm from which the cell undergoing meiosis originated (see below).

Second Meiotic Division (Meiosis II)

This division differs from mitosis and the first meiotic division in two ways (Fig.11-1): For one, it is not preceded by a new synthesis of DNA to form doubling-chromosomes. Instead, this division starts with cells having all of their chromosomes in the doubling-chromosome stage. Secondly, the dividing cells are haploid, not diploid, containing the haploid number of twenty-three doubling-chromosomes. As a consequence of these two features, the daughter cells that form will contain the haploid number of twenty-three chromosomes.

Thus, when the spindle develops in Meiosis II, all twenty-three doubling-chromosomes line up at the equator just as in mitosis, one doubling-chromosome per spindle fiber. They do not pair up in the second meiotic division as they did in Meiosis I because there are no homologous doubling-chromosomes to pair up with. In the second meiotic division, just before the doubling-chromosomes prepare to separate, the centromere of each doubling-chromosome is reproduced, and then the two sets of twenty-three separate chromosomes move toward each end of the spindle. Finally, each of the

two cells that form have the haploid number of twenty-three chromosomes (Fig.11-1).

Each of the two haploid cells formed from the second meiotic division of one cell having twenty-three doubling-chromosomes will have a complete set of twenty-three single chromosomes. Those two cells will have the identical chromosomes and genes except in the cases where crossover occurred in Meiosis I (see below).

Significance of Meiosis in Evolution

The diploid fertilized egg that results from the union of a haploid sperm and a haploid egg can differ genetically from either of its parents because of: (a) the *random mixing* of the chromosomes of maternal and paternal origin in the first meiotic division of pre-gamete cells that form eggs and sperm; (b) *crossing over* of the doubling-chromosomes in Meiosis I (see below); and (c) the mixing of the two sets of haploid chromosomes from the sperm and the egg that occurs during *fertilization.*

The greater the number (n) of chromosomes in the haploid set of a species, the greater the variety of chromosome mixtures of paternal and maternal origin that can occur in the gametes that form through meiosis.

For example, if n = 1, then the individual originally received one chromosome from the father and one from the mother. That individual, we say, has one set of homologous chromosomes. When that individual forms gametes, half will have the chromosome like the one given by the mother, and half will have a chromosome like that originally given by the father.

It gets more complicated if n = 2 as in Figure 11-1. The cells of that individual will have two sets of homologous chromosomes, that is two chromosomes from each parent. Assuming that no crossover takes place, the individual with n = 2 could have four kinds of distribution patterns of the chromosomes found in his or her gametes. The gametes could have (a) two chromosomes of maternal origin, (b) two of paternal origin, (c) chromosome number one of maternal origin and chromosome number two of paternal origin, and (d) chromosome number one of paternal origin and chromosome number two of maternal origin.

The possibilities become even greater the larger the number (n) of chromosomes in the gamete of a species. Because we are always dealing with pairs of homologous chromosomes in the diploid cell undergoing meiosis, a certain mathematical pattern prevails: the number of possible arrangements in the final haploid gamete equals two to the power which is equivalent to number of chromosomes in the haploid set. The formula is 2^n equals the number of possible arrangements.

For a species in which n = 1, there are 2^1 possible arrangements, or two. For a species in which n = 2, 3, or 4, there can be respectively four, eight, or sixteen arrangements. Let us take the supposedly simple example of a species in which n = 3 (Fig. 11-3). Note that a diploid pre-gamete cell of this species has six chromosomes, that is pairs of (a) large chromosomes with arms of equal length, (b) small chromosomes with arms of equal length, and (c) large chromosomes with arms of unequal length. The figure shows that during Meiosis I the pairs of homologous doubling-chromosomes arrange themselves so that two of the daughter cells formed after Meiosis II have a distribution of their chromosomes as in Figure 11-3(1), and the other two daughter cells formed during Meiosis II have a distribution as in Figure 11-3(2). Could the pairs of homologous doubling-chromosomes have arranged themselves differently in Stage II of Meiosis I so that different distributions of those six chromosomes could have occurred in the gametes finally formed after Meiosis II? The answer, as told by the formula 2^n, is obviously yes. There are six other possible distributions of chromosomes: (3) through (8) as illustrated in the column on the right in Figure 11-3.

In human beings n = 23. Thus, there are 2^{23} = 8,388,608 possible distributions of chromosomes in human gametes. Roughly speaking, that means in a human female there is only one chance out of eight million four

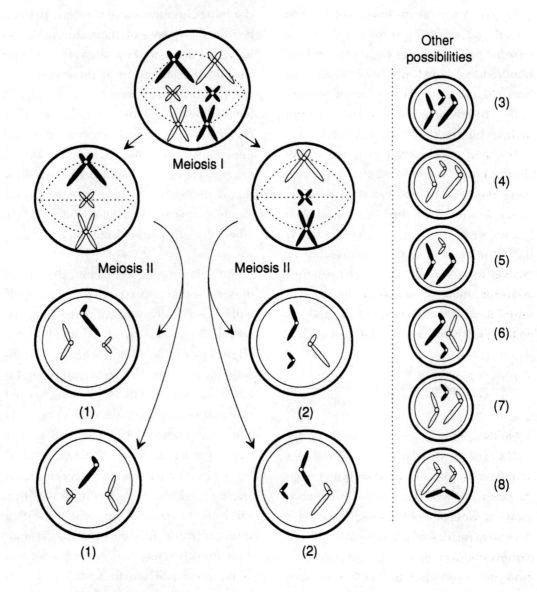

Figure 11 - 3 Possible distributions of chromosomes of maternal (black) and paternal (white) origins in gametes formed during meiosis from a pre-gamete cell having six chromosomes, i.e. n = 3.

hundred thousand that her egg will have the identical chromosomes as contributed by either the father or the mother of the individual producing the gametes. The same statistics applies to the distribution of chromosomes in the sperm produced by the male.

To carry these calculations further, when a man and woman procreate a child, the chances are one out of eight million four hundred thousand times eight million four hundred thousand, or one out of seventy million five hundred thousand that any two of their children (except identical twins) will have the identical combination of forty-six chromosomes. When we add to that probability the even further variation that can occur in crossover during meiosis (see below), we can see that meiosis provides for the evolution of the species by providing for genetic variability. In no uncertain terms, you must understand meiosis in order to understand sexual reproduction, and, as you will learn in Chapters 13-15, genetics.

Crossing Over

Another source of genetic variability originates in Stage III ("lining up") of the first meiotic division. Because during this stage the pairs of homologous doubling-chromosomes at the center of the spindle fibers are so close to each other, often equal lengths of chromatids from opposing doubling-chromosomes intertwine and exchange. This process is called *crossing over*.

Consider the case of a doubling-chromosome of maternal origin lining up next to its homologous doubling-chromosome of paternal origin. If small pieces of homologous doubling-chromosomes exchange equal and comparable pieces of their chromatids as indicated in Figure 11-4, then after the first meiotic division, as far as those doubling-chromosomes are concerned, one cell will have a doubling-chromosome containing a chromatid of maternal origin (black), and one of mostly maternal and some of paternal origin. Conversely, the other cell will show that doubling-chromosome to contain one chromatid of paternal origin (white), and the other of mostly paternal and some of maternal origin.

Thus, after the second meiotic division of those two cells each of the resultant four cells will be genetically different in the composition of the chromosomes that came from the original pair of doubling-chromosomes undergoing crossover during Meiosis I (Fig.11-4): (a) one gamete will have the unchanged maternal chromosome (black); (b) one will have that maternal chromosome with a piece of paternal chromosome substituted for a piece of the maternal chromosome (mostly black and some white); (c) another will have the paternal chromosome with a small part made up of the maternal chromosome (mostly white with some black); and (d) the fourth will have the unchanged chromosome of paternal origin (white).

It should be apparent that crossing over, even in its simplest form, allows for even more genetic variability than would occur if this process did not occur. I say simplest

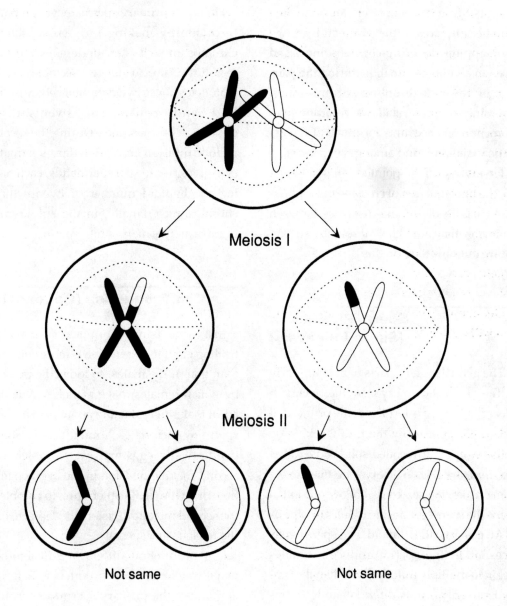

Meiosis I

Meiosis II

Not same   Not same

Figure 11 - 4 Simple case of crossing over.

form because crossing over can occur in a number of places in the chromatids of two overlapping doubling-chromosomes, and even in all four chromatids during the pairing of the two doubling-chromosomes. Crossing over is a relatively common phenomenon and accounts for much of the genetic variation found among the gametes.

Finally, as will be pointed out in Chapter 15, the phenomenon of crossing over allows biologists to determine distances between genes on the same chromosome. Can you figure out how they do this?

Sperm Production (Spermiogenesis)

The genital system starts to develop early in the life of the embryo. By the twentieth day after the egg is fertilized, the primordial germ cells appear. By the fifth week, these cells begin to form coiled solid rods, called *sex cords* (p. 271). In embryos destined to become males, the sex cords hollow out to become the sperm-producing tubules (Chapt. 2).

At puberty the diploid pre-sperm cells present in the sperm-producing tubules begin to multiply mitotically. When the diploid pre-sperm cells differentiate into diploid primary spermatocytes, the process of sperm formation is ready to begin. This entire process beginning with a primary spermatocyte takes about 64 days (Fig. 11-5).

First each primary spermatocyte prepares to divide by making forty-six doubling-chromosomes. It then undergoes the first meiotic division to give two secondary spermatocytes; these two daughter cells are haploid and consist of twenty-three doubling-chromosomes. During the second meiotic division each secondary spermatocyte gives rise to two spermatids, each having the haploid number of twenty three chromosomes. Finally, in the epididymis, the spermatids mature into sperm.

Egg Production (Oogenesis)

Just as the female hormonal system controlling gamete production is more complex than that of the males, so too is the process by which females make gametes. Whereas adult males manufacture about fifty million sperm by meiosis each minute, only about two to twelve eggs may ever complete the second meiotic division in the average lifetime of a woman. Furthermore, to complete the second meiotic division, the "egg" needs the nucleus of a sperm!

Egg production differs in still another major way from sperm production. In forming sperm, the primary spermatocytes undergo the first meiotic division immediately after the diploid number of doubling-chromosomes are made. Such is not the case with the production of eggs. In females, the pri-

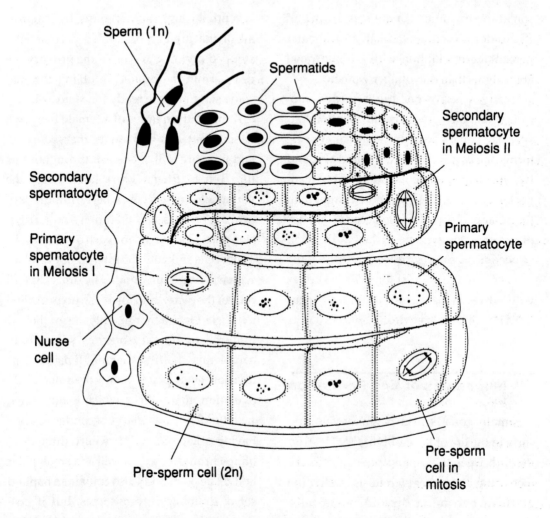

Figure 11 - 5 Schematic diagram of spermiogenesis within a sperm-producing tubule of an adult male. In this process one 2n pre-sperm cell becomes two 1n sperm.

mary oocytes and their forty-six doubling-chromosomes remain submerged in the ovary from twelve to fifty years before they undergo their first meiotic division to become secondary oocytes having the haploid number of twenty-three doubling-chromosomes. During those years, much can hap-

pen to those eggs that may complicate the meiotic divisions. In fact, the longer the interim between the development of the primary oocytes containing forty-six doubling chromosomes in the fetus and Meiosis I in the sexually mature female, the greater the chance for an egg to be formed with chromosomal abnormalities. For this reason, we find that older women give birth to a greater percentage of children with chromosomal aberrations than do younger women.

Finally, whereas one primary spermatocyte eventually gives rise to four haploid sperm of equal size, a single primary oocyte forms one large yolk-filled haploid egg, and two smaller haploid polar bodies which cannot be fertilized. The egg is much larger than the polar bodies and sperm because its cytoplasm contains extra nourishment that will be needed once the egg is fertilized and must travel down the oviduct for five to six days without receiving significant nourishment from its immediate environment.

Mechanics of Egg Formation

Female gonads also begin developing early in the life of the embryo (Fig. 11-6). By the fifth week the diploid pre-egg cells are increasing in number by mitosis, with some of them becoming diploid nurse cells. Groups of nurse cells cluster around a single pre-egg cell and provide it with nourishment. After the embryo is three months of age, some of the pre-egg cells become primary oocytes; these cells form the diploid number of forty six doubling-chromosomes in order to prepare for the first meiotic division. The formation of primary oocytes continues in the ovary of the embryo until the seventh through ninth months; by that time all of the pre-egg cells have permanently stopped dividing and all of the primary oocytes are in the diploid doubling-chromosome stage and are ready for Meiosis I.

At birth, the ovaries of a female baby will have about two million primary oocytes, and no more will ever form throughout its life. Ten to fifteen years later, when the young woman reaches puberty, hundreds of primary oocytes start their first meiotic division once a month during each ovarian cycle. These cells sink into the interior of the ovary where they are surrounded by other diploid cells of the ovary, and form a structure called a *follicle*. Once a month, only one of the few hundred of primary oocytes succeeds in completing the first meiotic division. This division takes place within the follicle. On ovulation, that follicle, which is now filled with fluid, releases a large secondary oocyte having a haploid set of twenty three doubling-chromosomes, as well as a small polar body. The polar body also contains a haploid set of doubling-chromosomes, but it possesses little cytoplasm. It may undergo Meiosis II, but, because it lacks sufficient cytoplasm, it cannot be fertilized and it usually disintegrates.

On ovulation, the single secondary oocyte, which is now in the oviduct, is ready for the second (final) meiotic division. Meiosis II, however, will not occur unless the secondary oocyte is first fertilized, i.e. accepts the

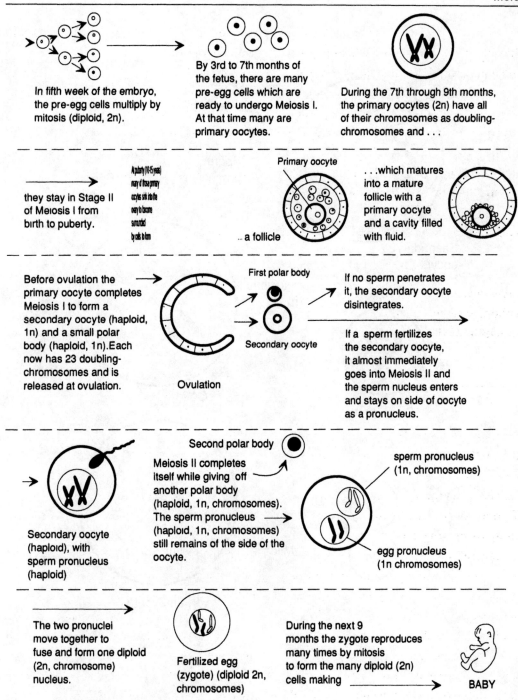

In fifth week of the embryo, the pre-egg cells multiply by mitosis (diploid, 2n).

By 3rd to 7th months of the fetus, there are many pre-egg cells which are ready to undergo Meiosis I. At that time many are primary oocytes.

During the 7th through 9th months, the primary oocytes (2n) have all of their chromosomes as doubling-chromosomes and . . .

they stay in Stage II of Meiosis I from birth to puberty.

At puberty (10-15 years) many of those primary oocytes sink into the ovary to become surrounded by cells to form . . .

. . . a follicle

Primary oocyte

. . . which matures into a mature follicle with a primary oocyte and a cavity filled with fluid.

Before ovulation the primary oocyte completes Meiosis I to form a secondary oocyte (haploid, 1n) and a small polar body (haploid, 1n). Each now has 23 doubling-chromosomes and is released at ovulation.

Ovulation

First polar body

Secondary oocyte

If no sperm penetrates it, the secondary oocyte disintegrates.

If a sperm fertilizes the secondary oocyte, it almost immediately goes into Meiosis II and the sperm nucleus enters and stays on side of oocyte as a pronucleus.

Secondary oocyte (haploid), with sperm pronucleus (haploid)

Second polar body

Meiosis II completes itself while giving off another polar body (haploid, 1n, chromosomes). The sperm pronucleus (haploid, 1n, chromosomes) still remains of the side of the oocyte.

sperm pronucleus (1n, chromosomes)

egg pronucleus (1n chromosomes)

The two pronuclei move together to fuse and form one diploid (2n, chromosome) nucleus.

Fertilized egg (zygote) (diploid 2n, chromosomes)

During the next 9 months the zygote reproduces many times by mitosis to form the many diploid (2n) cells making

BABY

Figure 11 - 6 Summary of stages in oogenesis.

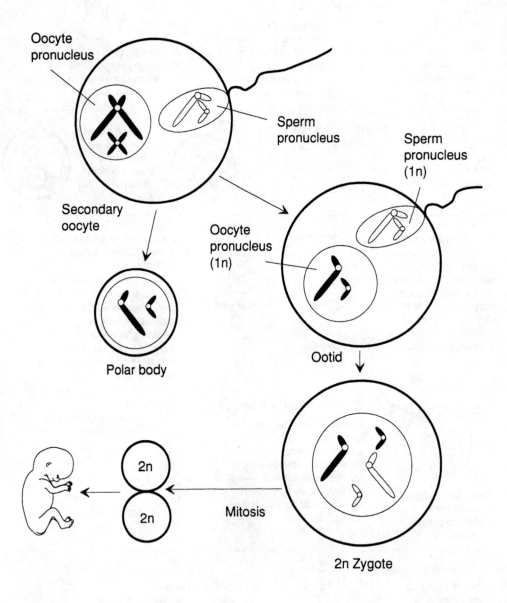

Figure 11 - 7 Chromosomes in egg, sperm, and zygote during fertilization.

haploid nucleus of a sperm (Chapt. 17). After the sperm nucleus enters, it stays to the side of the secondary oocyte and waits until that oocyte completes Meiosis II (Fig. 11-7). On completion of Meiosis II, a new haploid polar body is cast off, and the egg remains with its haploid number of chromosomes and also the former nucleus of the sperm. These two haploid nuclei meet near the center of the egg when they unite to give a diploid nucleus of forty six chromosomes; a new nuclear membrane then forms around them. The egg, now fertilized, is called a *zygote*. It starts to divide mitotically, and, in nine months, a child will usually be born.

Meiosis and Birth Defects

Of the three major categories of birth defects, most of those called chromosomal aberrancies originate from malfunctions of the movements of chromosomes that take place during the meiotic steps involved in the formation of eggs and sperm. These malfunctions and the resultant birth defects are subjects of the next chapter.

CHAPTER 12

Chromosomal Aberrancies

Cells with *chromosomal aberrancies* are defined as those having abnormalities in the structure and/or number of their chromosomes. Chromosomal aberrancies are known to be the underlying causes of a number of: (a) miscarriages that occur unnoticed, and (b) birth defects such as Down's syndrome. From our knowledge of meiosis, it becomes clear that there are many stages in the formation of sperm and eggs where errors in the distribution of the chromosomes can occur. Chromosomal aberrancies can also result from errors taking place during fertilization and from those occurring during the mitotic reproduction of cells in the early embryo. Let us consider some of those situations starting with the formation of gametes by meiosis and continuing through fertilization and the multiplication of cells in the very early embryo.

Chromosomal Aberrations Originating in Meiosis I

When pairs of homologous doubling-chromosomes come close to each other at the center of the spindle during Stage III of Meiosis I (Fig. 11-1), many possibilities arise for the chromatids of the opposing doubling-chromosomes to contact each other and interact. We have already discussed one of those possibilities, the phenomenon of crossing over (Fig. 11-4). Crossing over, however, does not lead to cells with abnormal chromosomes because equal parts of the intertangled opposing chromatids exchange places.

A number of aberrations, however, can occur to doubling-chromosomes undergoing difficulties with the pairing process of the first meiotic division. Two of these aberrations are deletions and translocations. The abnormal chromosomes formed by these processes are found in the gametes which are produced, and if these gametes achieve fertilization, the embryo will develop abnormally, if at all.

Deletions

A deletion refers to the processes by which a chromosome loses a piece of itself. The simplest form of a deletion occurs when a piece of a chromosome breaks off and is permanently lost from the nucleus. Deletions can also occur as a result of an unequal crossing over. During such an exchange, one chromatid of a doubling-chromosome gets a bigger piece of the chromatid of the opposing homologous doubling-chromosome. The one that got the smaller piece has a deletion, i.e. it lacks part of a chromatid, whereas the other chromatid actually has a duplication of part of that chromatid.

Most gametes which have a deletion of part of a chromosome, and which achieve fertilization, will never be expressed because the embryo will never develop. There are examples, however, of children being born with a deletion; children with one type of such a deletion are said to have the cri-du-chat ("cat cry") syndrome. In such cases, a piece of the number five chromosome is lost. These children have a number of malformations, are mentally defective, and, during infancy, cry in a manner sounding like a cat. Other cases of children with a chromosomal deletion show that parts of chromosome number eighteen are missing. These individuals reach adulthood, but are mentally retarded and have a small head and a short stature.

Another way that a deletion can arise is through a process called *translocation*.

Translocations

Translocations are chromosomal abnormalities which, when present in gametes, can lead to individuals who survive, but who also have developmental abnormalities. Translocations occur during Meiosis I when a chromatid from one doubling-chromosome breaks and a piece of it attaches to a chromatid of another doubling-chromosome which is not necessarily a homologous doubling-chromosome. As a consequence of a translocation taking place in a single primary oocyte or spermatocyte, not only will one gamete possess a chromosome with a translocated piece of another chromosome attached to it, but another gamete will show a chromosome with a *deletion* of the piece of the chromosome which was translocated. For a graphic reconstruction of translocation and a deletion taking place in two daughter cells of a meiotic division, look at Figure 12-1.

A relatively common translocation, in which a major portion of chromosome number twenty one attaches to chromosome number fourteen, leads to individuals having symptoms of Down's syndrome. It is not hard to see why symptoms of this syndrome occur. If the egg has all its twenty-three chromosomes, but an extra "dose" of chromosome number twenty-one attached to chromosome number fourteen, then the fertilized egg resulting from the union of that egg with a normal haploid sperm will have three "doses" of chromosome number twenty-one. One will come from the sperm, one from the egg, and "one" more from the egg because it has a piece of chromosome

twenty-one attached to chromosome number fourteen.

With newer and subtler techniques now available for investigating genes and chromosomes, more translocations and deletions are being discovered. One such aberrancy is called *Williams syndrome*. People with Williams syndrome show varying degrees of mental retardation and also remarkable abilities with language and music. Recent research (1993) demonstrates that they have a small deletion ("*microdeletion*") in one of their two chromosome 7s.

Translocations and Cancer

Some cancers can be correlated with individuals having chromosomes exhibiting specific translocations. For example, the cells of individuals showing retinoblastoma, a malignant tumor of the eye, are known to possess a chromosomal translocation. Likewise, a form of leukemia known as chronic myelocytic leukemia is correlated with a translocation between chromosomes nine and twenty-two.

Presumably, the new arrangements of the genes on the host chromosome and the piece of translocated chromosome activate a gene which can cause cancer. Normally this gene, called an oncogene ("onco" refers to cancer), is thought to be relatively inactive. See Chapter 16 regarding the activation of genes.

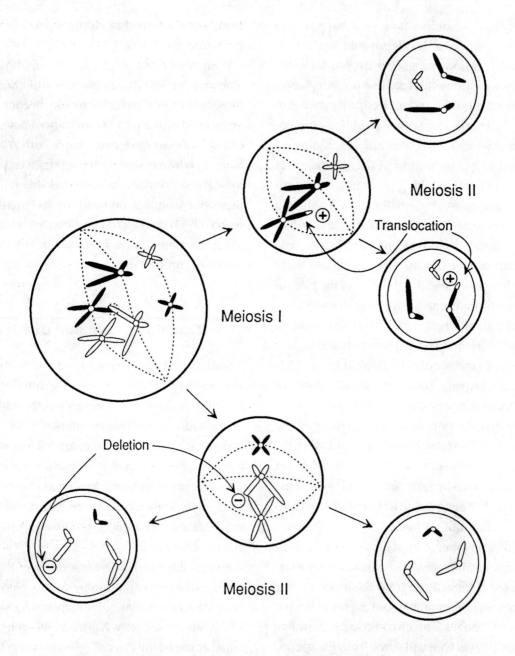

Meiosis II

Translocation

Meiosis I

Deletion

Meiosis II

Figure 12 - 1 Formation of chromosomes with a translocation (+) or a deletion (−).

Chromosomal Aberrations Originating with Errors Taking Place During Meiosis II

Errors of chromosomal movement taking place during Meiosis II can eventually lead to conditions known as a *trisomy* and a *monosomy*. A cell exhibiting monosomy is diploid, but lacks one chromosome; that is it contains 2n - 1 chromosomes in its nucleus, or in human beings, forty-five. We have already discussed such monosomic cells when we described the cells containing forty-five chromosomes in mosaics formed as a result of chromosome lag (Chapt. 10). A *trisomic cell*, on the other hand, has an extra chromosome. That is, its nucleus contains 2n + 1 chromosomes, or forty-seven in humans.

Nondisjunctions

Although chromosome lag can take place during Meiosis I and/or II (Fig. 12-2), and can lead to the production of a gamete lacking a chromosome, when the process known as nondisjunction takes place during the formation of either eggs or sperm, the gametes which are produced will either lack a chromosome or will possess an extra one. When those gametes form a zygote, the cells of the resultant embryo will demonstrate either a monosomy or a trisomy.

Nondisjunction refers to a process whereby the two chromatids in a doubling-chromosome do not separate (i.e. do not "disjoin") as they normally would during Stages III and IV of Meiosis II (Fig. 11-1). Consequently the nucleus of one of the two daughter gametes formed by that second meiotic division will contain two chromosomes from the "non-disjoined" doubling-chromosome, whereas the other daughter gamete developing after the nondisjunction in Meiosis II will be lacking a chromosome (Fig 12-3).

Note that the two chromatids of the doubling-chromosome that did not disjoin finally separated into two chromosomes, but only after that doubling-chromosome was enclosed within the nucleus of one of the daughter gametes (Fig. 12-3). The gamete with the extra chromosome now has n + 1 chromosomes, or twenty-four in humans. When that gamete gives rise to a fertilized egg, the cells of the child derived from that union will be *trisomic* (Fig. 12-4). Therefore, all the cells of the child will contain forty-seven chromosomes. We use the prefix "tri" because the zygote now has three of one type of chromosome rather than the normal number of two.

It is possible, but rare, for a cell to be trisomic for two chromosomes. In that case, the

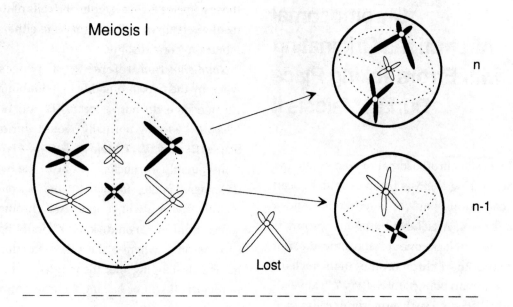

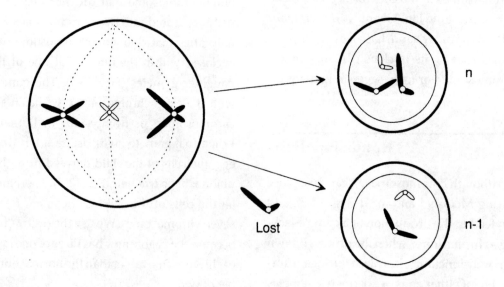

Figure 12 - 2 Chromosome lag in Meiosis I or II.

cell would have forty-eight chromosomes; one of the extra chromosomes would be the same as those in one pair of homologous chromosomes, and the other extra one the same as those in another pair of homologous chromosomes.

As stated above, the other daughter gamete cell formed by a nondisjunction in Meiosis II will lack one chromosome in its nucleus, and therefore, will contain n-1 chromosomes, or twenty-two in humans. Consequently, when that gamete forms a fertilized egg, the cells of the child derived from that union will be *monosomic* for the lost chromosome, i.e. they all will lack the missing chromosome, and will contain only

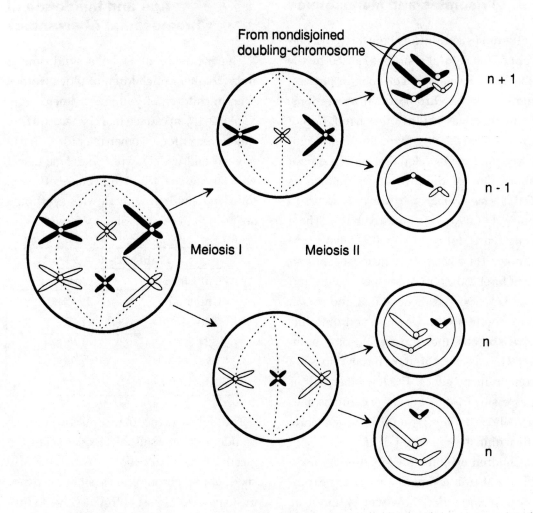

Figure 12 - 3 Case of nondisjunction during Meiosis II.

one chromosome of the type lost to the other gamete during nondisjunction (Fig. 12-4). It is also possible, but rare, for a cell to have forty-four chromosomes and be monosomic for two different chromosomes.

Incidences of Trisomies and Monosomies

Trisomies account for about forty-five percent of the natural abortions caused by the abnormal development of embryos possessing cells with chromosomal aberrations. Some trisomies, on the other hand, do not cause severe damage to the developing embryos, and the developing children do not abort. One of these trisomies, known as *Down's syndrome*, or *trisomy 21*, occurs in one out of five to six hundred births. There is no "cure" for children with Down's syndrome. They have a characteristic prominent forehead, short neck, open mouth, and low I.Q. Scientists believe that one reason why children with Down's syndrome survive is because the extra chromosome, number 21, is a small chromosome and does not contain many genes. The few other fetuses possessing trisomies of larger chromosomes usually do not survive until birth, or, if born, die within their first year of life.

Children exhibiting a monosomic disorder are also quite rare. One of the more common *monosomies*, however, known as

Turner's syndrome, arises when the cells of the child contain only one of the X chromosomes and no Y chromosome. I discuss Turner's syndrome more fully below in the section dealing with aberrations of the sex chromosomes.

Age and Incidence of Chromosomal Aberrancies

The incidence of Down's syndrome is more frequent in children of older women than in children of younger women. From Table 12-1, you can see that chances are fifty times greater for a woman over forty-five to have a child with Down's syndrome than a woman under thirty. An average of seven thousand children with Down's syndrome are born in the United States yearly.

Table 12 - 1	
Age of Mother	Frequency
Under 30	1/1,500
30-34	1/750
34-39	1/600
40 - 45	1/300
45 and older	1/30

The high incidence of trisomies in children of older mothers is usually explained by the fact that the primary oocytes, with their forty six doubling-chromosomes, sit in the ovary in suspended Meiosis I from twelve to fifty

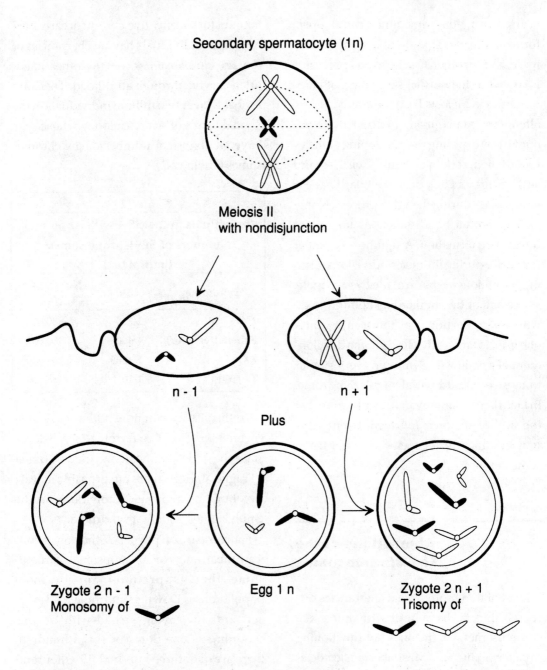

Figure 12 - 4 Nondisjunction in formation of sperm leading to zygotes possessing either a monosomy or a trisomy of one type of chromosome.

years. During that time much can happen: for example, the spindle fibers, the centromere, and/or other structures and processes involved in the normal separation of chromosomes in Meiosis II can be altered. These alterations can be made to occur more frequently by exposing the developing gametes to radiation, certain chemicals, viruses, and such drugs as LSD, or by cellular changes taking place during the aging process. Nonetheless, not all nondisjunctions take place during egg formation. A significant number take place during the meiotic divisions forming the billions of sperm made by each male.

The age of the mother has nothing to do with the production of a "fourteen/twenty-one translocation" that leads to the development of a child with symptoms very similar to those of a child with Down's Syndrome. In fact, there is some evidence indicating that the ability to form fourteen/twenty-one translocations during meiosis is inherited in some families.

Aberrations of the Sex Chromosomes

Most of the chromosomal aberrancies discussed thus far have dealt with autosomal chromosomes, i.e. the ones not functioning in determining sex. Most aberrancies dealing primarily with autosomal chromosomes, such as trisomies and triploids,

account for seventy five percent of early miscarriages. Individuals having aberrancies of the *sex chromosomes*, on the other hand, often survive through adulthood. There are probably over two million individuals in the United States of America today whose cells have an abnormal number of sex chromosomes (Table 12-2).

Table 12 - 2		
Individuals in the USA with Abnormal Numbers of Sex Chromosomes (partial list)		
Syndrome	Odds	Numbers in USA
XYY	1/250	850,000
Klinefelter (XXY)	1/400	530,000
XXX	1/1000	200,000
Turner (XO)	1/10,000	85,000

To this list we can now add a newly discovered *Fragile X* syndrome (see below); since the odds are one in one thousand that a Fragile X arises, there are probably at least two hundred thousand individuals in the United States with that syndrome.

For comparison purposes, the only significant aberrancy of an autosomal chromosome that is predominant in most populations is Down's syndrome. The average odds for having a child with Down's syndrome is one in six or seven hundred; there are over three hundred fifty thousand individuals with Down's syndrome in the United States.

Sex Determination

The major factor determining the sex of the individual is the presence or absence of a single Y chromosome. Those individuals who have no Y chromosome, but instead have two X chromosomes, are normal females, i.e. they are XX. Individuals who have a Y chromosome as well as an X chromosome, are normal males, i.e. they are XY. Even individuals that exhibit the chromosomal aberrancy of XXXXY are males, although they are not normal in all respects. Likewise, individuals having no Y chromosome and more or less than two X chromosomes are female, though not perfectly normal in all respects. We can also say that the sex of an individual is determined at conception when a sperm nucleus containing either a Y or an X chromosome combines with the nucleus of the egg which always contains an X chromosome. The genetic, hormonal, and developmental bases of sex determination are described in Chapter 22.

Aberrations in the Number of Sex Chromosomes

Most aberrations in the number of the sex chromosomes in the cells of an individual owe their origin to either (a) nondisjunctions of those chromosomes during the production of one of the gametes used to form the fertilized egg, or to (b) chromosome lag of a sex chromosome in early mitotic divisions of the fertilized egg.

XO, Turner's Syndrome

The monosomic condition known as *Turner's syndrome* can develop when an egg lacking an X chromosome because of a nondisjunction is fertilized by a sperm bearing an X chromosome (Fig. 12-5). Children with Turner's syndrome have a characteristic wide neck and small stature. After puberty they possess rudimentary ovaries and underdeveloped breasts. They do not menstruate or ovulate. One out of every three thousand five hundred children born have Turner's syndrome. There does not appear to be any relationship between the incidence of Turner's syndrome and the age of the mother.

An XO combination, i.e. forty four autosomal chromosomes and one X chromosome, usually arises from a fertilization taking place between an egg (or sperm) lacking an X chromosome through nondisjunction or chromosome lag, with a sperm (or egg) bearing an X chromosome (Fig. 12-5).

About twenty percent of natural abortuses are XO embryos. As indicated before, women who are XO and who survive until adulthood do not ovulate or menstruate. Their ovaries do not develop fully. Thus, one

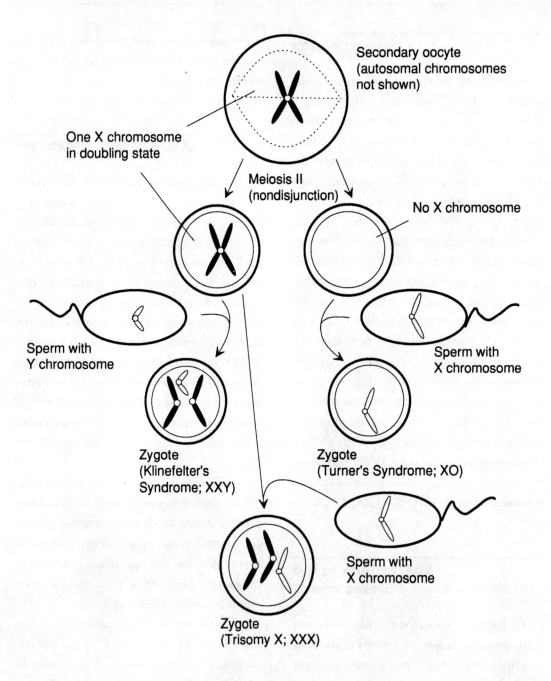

Secondary oocyte
(autosomal chromosomes
not shown)

One X chromosome
in doubling state

Meiosis II
(nondisjunction)

No X chromosome

Sperm with
Y chromosome

Sperm with
X chromosome

Zygote
(Klinefelter's
Syndrome; XXY)

Zygote
(Turner's Syndrome; XO)

Sperm with
X chromosome

Zygote
(Trisomy X; XXX)

Figure 12 - 5 Nondisjunction of X chromosome in egg can lead to zygotes showing Klinefelter's syndrome, Turner's syndrome, or Trisomy X.

X chromosome is sufficient to initiate ovary development, but two X's are needed for the ovaries to complete development and function normally.

XXY, Klinefelter's Syndrome

Because individuals with Klinefelter's syndrome have a Y chromosome, they have predominantly male features. Yet the presence of an extra X chromosome affects their development negatively; they have small testes which do not produce sperm, and some of these individuals develop breasts. Often they are mentally retarded. Like women with Turner's syndrome, men with Klinefelter's syndrome can originate from an egg that went through nondisjunction of the X chromosome. In the case of Klinefelter's syndrome, two of the X chromosomes were retained in the egg which then combined with a normal haploid sperm having one Y chromosome (Fig. 12-5).

XXX, "Super Female"

The same mechanism of nondisjunction of an X chromosome that can lead to Klinefelter's syndrome, can also lead to the production of a fertilized egg that is XXX. In this case, an XX egg or sperm which has gone through a nondisjunction of the X chromosome will combine with an X-bearing sperm (or egg) (Fig. 12-5). Women whose cells are XXX vary greatly in their mental and sexual capacities. Some are of normal intelligence, and others are retarded. Some are fertile and produce normal children, whereas the majority are infertile.

XYY Individuals

As indicated in Table 12-2, one out of every two hundred fifty children born may be XYY. Such situations can arise from the fertilization of an egg with a sperm that formed from the nondisjunction of a Y chromosome in the second meiotic division. Men who show this syndrome are taller on the average than normal XY males. They are usually mentally retarded, but not necessarily. Because many studies have been done on XYY individuals who are in prisons, there was a feeling among some medical scientists that the XYY syndrome is correlated with a tendency for criminal behavior and aggressiveness. Others are withholding judgement until more studies are carried out on XYY men who are not institutionalized.

Polysomies of Sex Chromosomes and Mosaics

There are a significant number of reports regarding individuals having polysomies of their sex chromosomes, i.e. they have more than two sex chromosomes. There are XXXX females; all are severely retarded mentally, but appear normal physically. Some males have been shown to be XXXY. Others are XXXXY, and still others have been shown to be XXYY. Individuals with these three male karyotypes have varying degrees of mental and physical handicaps.

In the case of one mosaic that was reported, the individual had XX, XXY and XXXYY chromosomes in his white blood cells, XX and XXY cells in his testes, and XX in some of his skin cells.

Aberrations Due to an Alteration of the X Chromosome

A newly discovered syndrome which is responsible for severe mental retardation in males, has been found associated with an abnormal X chromosome called the *Fragile X* chromosome. Analyses of the karyotype of institutionalized patients show that mental retardation associated with the Fragile X chromosome ranks second to that of Down's syndrome. Each year about five hundred male babies with the Fragile X syndrome are born in the United States. Interestingly, males affected with Fragile X syndrome appear to be normal in other respects. Some scientists think that there may be cases in which the presence of a Fragile X chromosome in males is linked to autism. Others suggest that some females having one Fragile X chromosome may have mild mental retardation. For these and other reasons, geneticists propose that the Fragile X chromosome affects brain development in some negative way.

The Fragile X chromosome is so named because one end appears pinched and, therefore, has a constriction with one small mass of chromosomal material at its tip (Fig. 12-6).

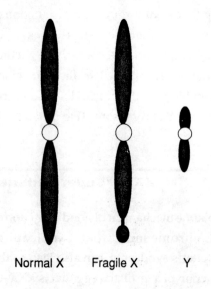

Normal X   Fragile X   Y

Figure 12 - 6 Sex chromosomes.

Sometimes the tip breaks off at the constriction and the Fragile X chromosome cannot be detected in karyotypes. When a Fragile X chromosome is paired with a normal X chromosome in a female, the effects of the fragile end will usually not be seen in that female. Yet that female (X, Fragile X) may pass on her Fragile X chromosome to her offspring; thus, she is called a *Fragile X carrier*. If her egg carrying a Fragile X chromosome unites with a sperm bearing a Y chromosome, the resultant male will exhibit the Fragile X syndrome. Why? Because the small Y chromosome cannot compensate for the genetic damage in the Fragile X chromosome as would a normal X chromosome. This association of developmental abnormalities caused by the combination of a defective X chromosome with a normal Y chromosome is discussed more fully under sex-linked (i.e. X-linked) genetic diseases (Chapt. 15).

Hermaphroditism

Hermaphrodites are named after the son of the Greek god Hermes and the goddess Aphrodite. As a teenager his body was fused with the body of a female who wished to possess him. Thus, hermaphrodites are supposed to have the reproductive capacities of both sexes, as do many lower organisms such as earthworms and some hydra. Among humans, however, the term hermaphrodite is usually applied to individuals who possess physical characteristics to varying degrees of both male and female genitalia. Some of these individuals show abnormalities of the sex chromosomes and others do not. Medical scientists describe a number of categories of hermaphrodites.

True hermaphrodites have both testicular and ovarian tissue. Sometimes they exist as separate organs, one on each side of the body. Other times they exist as one organ exhibiting properties of both the testis and ovary. Of the one hundred or so cases reported, many are XX/XY mosaics. Such individuals cannot function both as male and female, and for the most part cannot act effectively as either one.

Pseudohermaphrodites

Pseudohermaphrodites, in contrast to the so-called true hermaphrodites, have gonads of only one sex, but their sexual organs resemble to varying degrees that of the other sex. In males, we refer to pseudohermaphrodites as exhibiting *testicular feminization*. I say males, because karyotype analyses show most of them to be XY. These individuals exhibit external female genitalia, and yet may have undescended testes in their abdominal cavity. The vagina ends blindly with no uterus. There are no ovaries, but there are often well developed breasts. Presumably the female steroid hormones are produced by the undescended testes. The causes of testicular feminization are now better understood and are explained more fully in Chapter 22.

There are also a number of different kinds of female pseudohermaphrodites. These individuals are genetically XX, but exhibit varying degrees of maleness. More male hormones are produced than normally, and the clitoris is often greatly enlarged. Yet, female hermaphrodites possess ovarian rather than testicular tissue.

In recent years scientists have uncovered cases in which children who were thought to be girls "suddenly" developed into functional males at puberty. The underlying basis of this type of transitional hermaphroditism is also described in Chapter 22.

Sex Determination

For varying social, family, and medical reasons, prospective parents often exhibit if not concern, at least curiosity about the sex of an expected child. On the average, for every one hundred newborn girls, one hundred ten boys are born.

You have already learned that to be a normal female or male, an individual must be either XX or XY respectively with regards to the kinds of sex chromosomes within their cells. Based upon these facts, some scientists believe that the predominance in number of males born may have something to do with the fact that a sperm bearing a Y chromosome may have an edge on a sperm bearing an X chromosome in reaching the egg because the X chromosome is so much larger than the Y one. Thus, the sperm bearing the Y chromosome would be smaller and lighter and could swim to the egg faster.

Regardless of the social and family reasons for knowing the sex of the child beforehand, there may be some critical life-determining reasons for knowing whether or not the expectant child will be male. An example might be a situation in which a mother is known to be a carrier of a Fragile X chromosome. If her expected child is shown early in pregnancy to be male, then the chances are fifty/fifty that the boy will exhibit the Fragile X syndrome and be mentally retarded. The chances are fifty/fifty because there is an equal chance that the Y chromosome from the sperm is matched with the carrier mother's normal X chromosome in the egg; in such a case the boy would be normal. Thus, if the parents are certain that their expectant child will be male, they must decide whether to risk the chance of having a Fragile X son, or to choose to abort the embryo. These are very difficult decisions to make, and they depend on accurate methods for determining the sex while the expected child is still in the embryonic phase. In Chapter 14 on genetic diseases, you will learn much about a number of diseases, such as hemophilia, and color blindness, associated with boys who are children of mothers carrying genetic defects on one of their X chromosomes.

Barr Body Method for Determining the Number of X Chromosomes Present In a Cell

Although the sex of an individual can be detected by examining the karyotypes for the XX or XY conditions, there is a much simpler method. This method, called the *Barr body* test after its discoverer, is based on a finding that somatic cells of females before mitosis show a darkly stained body along the edge of the nucleus (Fig. 12-7). Because these Barr bodies were not seen in nor-

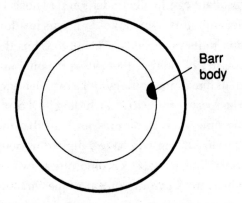

Figure 12 - 7 Somatic cell taken from a female showing one Barr body on the side of the nucleus.

mal males, their presence was considered a sign that those cells came from females.

From studies of cells from individuals of all kinds of karyotypes, it has been shown that the number of Barr bodies is always one less than the number of X chromosomes in the cell examined. Thus, a cell from a normal female with two X chromosomes will have one Barr body. A cell from a male (XY) will have no Barr body. Thus, under most situations, the presence of one Barr body indicates femaleness, and no Barr body indicates maleness.

This conclusion, however, does not always hold with cells from individuals with an abnormal number of sex chromosomes. For example, a cell from a male with Klinefelter's syndrome (XXY) would have one Barr body, whereas a cell from a female with Turner's syndrome (XO) would have none. Accordingly an XXX female would have two Barr bodies, an XYY male would have none, etc.

Each Barr body appears to be an inactive compact form of the "extra" X chromosomes in the cell. We are not certain of the reason why the extra X chromosomes are inactivated in resting somatic cells.

Not all somatic cells in normal XX females show a Barr body. The oogonia and the primary oocytes all need to have both X chromosomes in the active form. If they did not, meiosis could not take place. Also, both X chromosomes are needed early in the embryonic stage for the ovary to develop normally. Recall that cells of females having Turner's syndrome possess one X chromosome (XO) and do not develop functional ovaries.

Getting back to our example of an expectant mother who is a carrier of a Fragile X chromosome, she would know if her expectant child was a male if the cells taken from the amnion (see below) had no Barr body.

Errors Occurring During Fertilization

In Chapter 11 you read that forty percent of the embryos which abort naturally in the first three months of pregnancy have chromosomal aberrancies. Of those, twenty-three percent are triploids and five percent are tetraploids. *Triploid cells* have three haploid sets of chromosomes, or, in humans, sixty nine in all; human cells that are tetraploid contain four haploid sets of chromosomes, or ninety two. In general terms, triploids are 3n and tetraploids 4n. Such cells will have difficulty reproducing mitotically because of the large number of extra chromosomes in the nucleus; consequently the embryo will die very early in pregnancy and most likely will not implant (Chapt. 18) into the endometrium of the uterus.

Triploids can form in a number of ways. The most common is for the nucleus of the last polar body, i.e. the one formed in Meiosis II, to stay with the egg and to fuse with the pronuclei of the egg and sperm. Triploids can also form when two sperm happen to enter an egg at the exact same moment.

Tetraploids can form, for example, when two sperm enter an egg that has not discarded its polar body, and all four nuclei fuse. Can you think of other ways for triploid and tetraploid zygotes to form?

Formation of Hydatidiform Mole Aberrancy at Fertilization

One highly unusual aberrancy, called *hydatidiform mole*, can occasionally occur when a sperm nucleus containing an X chromosome fertilizes an egg. In those cases the resultant egg, while undergoing Meiosis II, gets rid of its own haploid nucleus in addition to its polar body (Fig. 12-8). Thus the sperm's haploid pronucleus is the only nucleus remaining in the cytoplasm of that egg. The sperm pronucleus, including its X chromosome, then duplicates itself so that the "fertilized" egg will have a diploid nucleus containing forty-six chromosomes, two of which are X chromosomes; all the chromosomes, however, originated from the nucleus of the sperm.

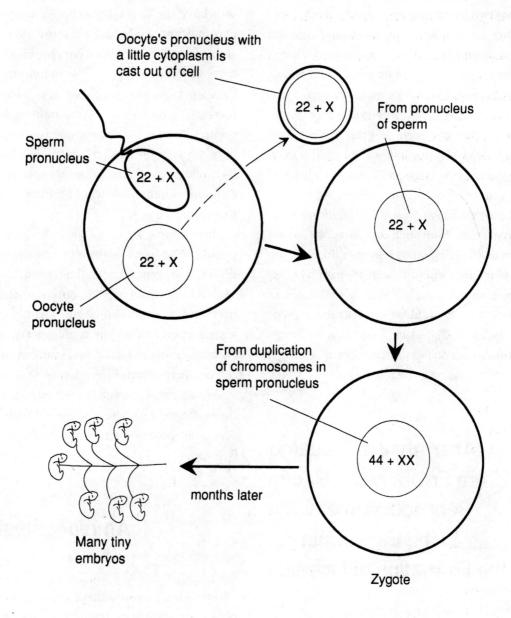

Oocyte's pronucleus with
a little cytoplasm is
cast out of cell

22 + X

From pronucleus
of sperm

Sperm
pronucleus

22 + X

22 + X

Oocyte
pronucleus

22 + X

From duplication
of chromosomes in
sperm pronucleus

44 + XX

months later

Many tiny
embryos

Zygote

Figure 12 - 8 Formation of a hydatidiform mole

You would think that such an egg, which has two X chromosomes, would survive and that the resultant embryo would develop into a female, but it does not. Instead, the egg starts to multiply, but its cells do not form an embryo. Rather they form an abnormal placenta having a mass of cysts that looks somewhat like a bunch of grapes with each "grape" becoming a tiny embryo. The mass grows quite large until it is expelled at around the twentieth week. In some cases the woman bearing this hydatidiform mole may suffer from such complications as toxemia, bleeding, and, infrequently, a highly malignant form of cancer. Women who experience a hydatidiform mole are known to bear normal children in subsequent pregnancies. It should be noted that hydatidiform moles do not form if a sperm bearing a Y chromosome fertilizes an egg.

Aberrancies Resulting from Errors in the Mitotic Reproduction of Cells in the Early Embryo: the Formation of Mosaics

If *chromosome lag* (Chapt. 10) takes place during the first cell division of a fertilized egg, half of all the subsequent cells that form will have the normal number of forty-six chromosomes, whereas all those derived from the cell that underwent chromosome lag will have forty-five. The individual to develop from that fertilized egg will be, therefore, a mosaic. If chromosome lag had occurred later in the development of the embryo, say after the tenth cell division or later, then obviously a smaller proportion of the cells of that individual would contain forty-five chromosomes.

Chromosome lag is not the only cause of mosaics. They can also develop when a *nondisjunction* occurs during the reproduction of cells in the embryo. With nondisjunction, in contrast to chromosome lag, not only will a proportion of the cells contain forty-five chromosomes, but an equal number will contain forty-seven chromosomes.

Mosaics can also result from aberrancies in the number of sex chromosomes, such as occurs in the case of true hermaphrodites (see above).

Amniocentesis

Through the procedure of *amniocentesis* physicians are able to obtain cells of the early embryo to examine for karyotype. First the physician inserts a needle into the sack of

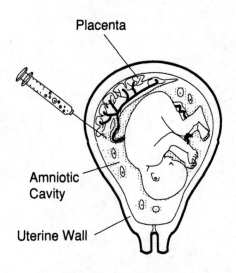

Placenta

Amniotic
Cavity

Uterine Wall

Figure 12 - 9 Amniocentesis is a technique for prenatal diagnosis that is carried out by inserting a sterile needle into the amniotic cavity and withdrawing a small amount of fluid for analysis.

fluid surrounding the embryo (amniotic sac) and removes a little of the fluid (Fig. 12-9). In the fluid there are usually a number of loose cells that have separated naturally from the embryo. These, however, are not sufficient to provide for a karyotype analysis because the cells must be stopped in the middle of mitosis not when the doubling-chromosomes line up. To get these embryonic cells to multiply mitotically, medical technicians must place them in test tubes containing nutrients. It usually takes nearly two weeks before the technicians can get a sufficient number of embryonic cells undergoing mitosis in order to make a reliable analysis of their karyotype.

The earlier that amniotic fluids can be withdrawn the better. Usually amniocentesis is done between the twelfth and sixteenth weeks of pregnancy. Before the twelfth week there is usually not enough amniotic fluid to withdraw safely. The sixteenth week is about the latest time to withdraw fluids because of the limit usually put on having abortions after the twenty fourth week. The physician must consider not only how much fluid is present in the amnion at the sixteenth week, but also that two to five weeks are needed to grow the cells needed for some of the tests to be performed on those cells.

Amniocentesis is usually recommended for women having children late in life, for women for whom there is a familial history of carrying a chromosomal aberrancy or genetic disease that can be analyzed, and for women known to be carriers of X-linked diseases. Amniocentesis can detect the sex of the embryo, chromosomal aberrancies, and many biochemical genetic and embryonic defects.

There are, however, some risks involved in the use of amniocentesis. Although these risks are considered to be ever so slight, they must be considered. Occasionally the fetus is damaged sufficiently so that it becomes infected, aborts, or both. There is a little risk of hemorrhaging, although it is not too common. Finally, one report indicates that if too much amniotic fluid is removed, the development of the brain is affected lowering the I.Q. of the child about 25 units.

In recent years a new test has been under development that eventually may prove to be a convenient, safe, and economical substitute for amniocentesis. This test, called *chorionic villi sampling, or CVS* for short, involves the physician removing a small bit of the placenta which surrounds the embryo and which is embedded in the wall of the uterus. Because the cells of the placenta have the same chromosomes as are present in the embryo, it is possible to examine their karyotype and therefore know the karyotype of the cells of the embryo (see Chapter 22).

CHAPTER 13

Genetics - the Basics

In this chapter and the next three you will be introduced to four aspects of human heredity: (1) First we will examine the basic principles of genetics as discovered through the revolutionary experiments in the mid-nineteenth century of Gregor Mendel working with pea plants. (2) After that we will relate Mendel's discoveries to some visible human traits and to some human genetic diseases that are controlled by one gene (Chapter 14). (3) Following that, Chapter 15 will cover five different kinds of situations in genetics in which two or more genes are involved. (4) Finally, in Chapter 16, we will learn how the principles of genetics are governed by the laws of molecular biology.

Although this chapter is the first one directed to the science of genetics, that is the study of how traits are passed on from one generation to the next, we have already set the stage for this study in the chapters dealing with meiosis. In fact we shall see that Mendel's two major laws as well as the process of *recombination* of reciprocal pieces of maternally and paternally contributed chromosomes can be explained by knowledge of meiosis.

We have already been introduced, for example, to the concepts of *chromosomes, DNA, alleles, homologous chromosomes,* and *crossing over.* Put in another way, we have learned that the cell's genes are composed of DNA covered by protective proteins, and that they are located on the chromosomes. Furthermore, we know that through meiosis each gamete gets one set of 23 chromosomes, and that each chromosome contains thousands of genes. At fertilization, two sets of 23 chromosomes, one set from the egg and one set from the sperm, form a single nucleus having 46 chromosomes. Each of the 23 chromosomes in that nucleus which came from the egg is matched with a corresponding chromosome of the same size, shape, and genetic information that came from the sperm. We called these *homologous chromosomes.* Finally, we have already learned that for each gene on a chromosome, there exists a corresponding gene, called its *allele,* on the chromosome homologous to it. For our purpose, we will call these two genes an *allele set.*

If you can recall these facts, can master a few more definitions (such as *dominant, recessive, genotype, phenotype, homozygous,* and *heterozygous*) and can learn a few simple mathematical operations, you will be able to gain a practical understanding of the basics of human genetics. A key point to keep in mind always is that whenever Mendel refers to a pair of factors affecting a trait, he in reality is referring to an allele set of genes on one pair of homologous chromosomes (see Fig. 13-1).

Mendelian Genetics

In the dozen years spanning 1856 through 1868, Gregor Mendel, an obscure monk in Poland, discovered the foundations of the science of genetics while observing various traits of pea plants in his garden. Because his findings were so far ahead of his time, because biologists had no understanding of meiosis at that time, and perhaps because his findings were published in a Polish journal not widely read internationally, Mendel's discoveries were not recognized until the beginning of the twentieth century. At that time the scientific world was ready for Mendel's discoveries, and three scientists almost simultaneously rediscovered Mendel's works.

Mendel was fortunate to have been working with pea plants because they offer a number of advantages for genetic research. For one, the flowers (the sex organ of the plant) usually *self-fertilize* the *ovule* (the plant's egg) with its own *pollen* (the plant's sperm) from the same flower; consequently, after many generations of this repeated self-fertilization, Mendel obtained many relatively pure genetic strains of pea plants. For example one strain had only red flowers whereas another had only white flowers.

Secondly, Mendel was able experimentally to *cross-fertilize* the pea plants by taking pollen from one strain of pea plant, say a white one, to fertilize the ovule of another strain of pea plant, say a red one. Thus, when Mendel examined the progeny from his artificial matings, called crosses, he was able to see with his own eyes whether or not those progeny retained the traits of the parental plant whose ovule was fertilized, or whether they gained some new characteristics transferred from the plant that donated the pollen.

Some Basic
Terms And Concepts

Before delving into Mendel's experiments, you need to become familiar with certain terms and concepts. The first, *phenotype*, refers to an observed characteristic, or *trait* of an organism. For example, if we were discussing the color of the flower of the pea plant, we would say that its phenotype is red if it appears red to the eye, or white if it appears white. But as Mendelian genetics will tell you, plants having the same visible traits do not always have the same genetic make-up, i.e. the same *genotype*. For example, as you soon shall read, a number of pea

plants having flowers with a red phenotype may differ in the genes making up the allele set which controls that red color; that is, those plants having red flowers may have a different genotype.

We now know that a particular *genotype* refers to a specific allele set of genes on a specific pair of homologous chromosomes. For example, the set of genes controlling red or white color of the pea plant flower can be said to have a genotype of RR, Rr or rr. Let us take the genotype (allele set) Rr as an example, and call the gene designated by a capital letter the *dominant* one, and the gene designated by the small letter as the *recessive* one. Figure 1 shows a sketch of six of the chromosomes of a pea plant. In that figure, and in subsequent ones like it, let us say that the dark colored chromosomes originally came from the ovule (female), whereas the colorless chromosomes originally came from the pollen (male). You can see that the alleles R and r are on their respective homologous chromosomes, the black and colorless large ones. Whenever the situation arises that at least one of the genes in the allele set under consideration is *dominant*, then the dominant characteristic will be expressed in the phenotype of the plant. For the *recessive* characteristic to be expressed, the genotype must be such that both the genes in the allele set are recessive, such as in the allele set rr.

Thus, we come to our final two new terms: *homozygous* and *heterozygous*, terms that

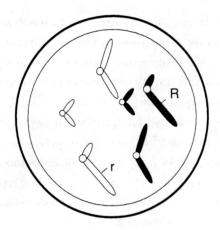

Figure 13-1 Three pairs of homologous chromosomes of pea plant showing one of the genes of the allele set Rr on a chromosome of female origin, and the other on the homologous chromosome of male origin.

apply to the genotype of an allele set. If both the genes of the allele set are the same, they are said to be homozygous; hence, the alleles can be homozygous dominant, as in RR, or homozygous recessive, as in rr. In contrast, the heterozygous state refers only to an allele set in which the gene on one chromosome is dominant, and the other gene on the other chromosome, called its allele, is recessive, as in the genotype Rr. Once you understand these terms as well as meiosis, then you should find Mendelian genetics relatively easy to comprehend.

Mendel's Initial Experiments:

Like his predecessors, Mendel found nothing unusual in learning that pea plants having red flowers crossed to others having red flowers produced seeds that became pea plants all having red flowers, and that plants with white flowers crossed with themselves or with other pea plants having white flowers, produced seeds that became pea plants all having white flowers. In shorthand, we would write:

Red   X   Red = Red

and

White   X   White = White

All these observations refer to traits that Mendel could actually see, i.e. the red and white phenotypes.

Mendel was puzzled, however, when he crossed a pea plant having red flowers with pollen from a plant having white flowers. He found that all of the plants coming from the seeds of such a cross had red flowers. That is:

Red   X   White = Red
 ♀   ♂

The symbol   refers to plants with an ovule ("egg"), and the symbol   to plants which contributed the pollen ("sperm").

Furthermore, Mendel took the plants with red flowers coming from that cross between plants with red flowers and plants with white flowers, and allowed them to self fertilize. To his surprise he found that 25% of the progeny of that self-fertilization had white flowers, and 75% had red flowers. We would write:

Red X Red =   25% white plants and
75% red plants

Geneticists have devised a special kind of shorthand, and would summarize the two crosses I have just described as follows:

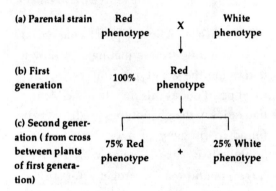

From his experiments, Mendel made the following revolutionary deductions: (1) that particles (or factors) coming from each parent, control the phenotype of each trait; (2) that these factors exist in pairs; and (3) these factors can be either dominant or recessive in their ability to allow a particular phenotype to be expressed.

We now know that (1) these factors are the genes contained in the chromosomes; (2) the pair of factors controlling a particular phenotype refers to the specific two genes in a

allele set; and (3) the dominant factor is a gene which is responsible for the production of a specific protein that will always be expressed in the phenotype, even in the presence of the protein controlled by the recessive gene, i.e. either in the homozygous dominant or heterozygous situation. The recessive gene, however, can be expressed in the phenotype only if it exists in the homozygous condition, that is as both genes of an allele set.

In the case of the red and white color of the flowers of Mendel's pea plants, the protein controlled by gene R leads to the production of a red pigment in the petals of the flower. One R, as in the heterozygous condition Rr, is sufficient to make enough of the correct protein to produce the necessary amount of red pigment. The homozygous dominant allele set RR will lead to the production of twice as much of that protein even though one dose is sufficient to give the plant a red color. On the other hand, the protein produced by the genes of a plant that is homozygous recessive, i.e. rr, is unable to make the correct protein which leads to the production of the red pigment. Hence, in that case, with no red pigment being produced, the petals remain colorless, or white.

With this information let us again analyze Mendel's experiments, but this time using the genotype shorthand.

We can verbalize this shorthand by saying that a cross of a parent plant which is

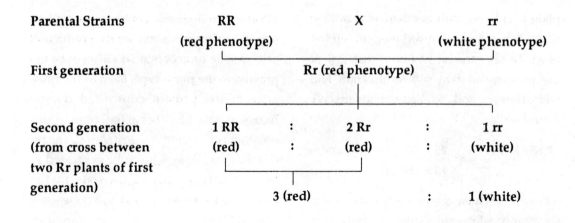

Parental Strains	RR		X		rr
	(red phenotype)				(white phenotype)
First generation			Rr (red phenotype)		
Second generation	1 RR	:	2 Rr	:	1 rr
(from cross between	(red)	:	(red)	:	(white)
two Rr plants of first					
generation)		3 (red)		:	1 (white)

homozygous dominant for red flowers (RR), with a parent plant which is homozygous for white flowers (rr) will give rise to a first generation of plants that phenotypically are all red, but genotypically are all heterozygous in reference to the genes R and r, or in the shorthand, Rr. Then, when we interbreed those first generation plants which are heterozygous (Rr) for flower color with each other, they will produce a population of second generation plants in which plants with red flowers outnumber plants with white flowers 3:1. An analysis of the genotype of the red plants in this second generation would show that for every one that was homozygous dominant (RR), there would be two other red plants that were heterozygous (Rr). All the plants with white flowers, however, were homozygous recessive (i.e., rr) for that allele set and contained no dominant R gene.

Populations

Before we learn how to predict the phenotypes resulting from a mating, and how to determine the genotype of an organism, I must point out at this time that the Mendelian ratios of the phenotypes predicted to be found in progeny of a particular mating apply only if we are dealing with relatively *large populations* of progeny (offspring), not with one, two or three progeny such as we often encounter in human families. In other words, Mendelian genetics deals with *statistical events*, with the chance that such and such a genotype will occur.

For example, in the illustrations given thus far, a cross between a plant that is homozygous dominant (RR) with a plant that is homozygous recessive (rr) will give first generation progeny, 100% of which are heterozygous (Rr) and, therefore, red. In the case, however, in which a heterozygous Rr plant is crossed with another Rr plant, the color of the flowers of the progeny do not necessarily fit the 3:1 ratio of red to white phenotype predicted by Mendelian genetics if only a small number of progeny develop. For example, from such an Rr x Rr cross, the first plant to blossom from seeds of that cross may have either white or red flowers.

Likewise, the second plant to develop could also have either white or red flowers, and so on and so on. If only four seeds were selected at random from that cross, then possibly all four seeds would develop into plants having red flowers, or, on the other hand, even into plants all having white flowers. However, the greater the number of seeds from that cross which germinated, then the greater the chance for the 3 red to 1 white ratio to occur.

It is almost like flipping coins. You know that if you flip a coin 1,000 times, about 500 times you will get heads coming up, and tails the other 500 times. Yet, if you flip a coin a few times only it is possible to get either heads all the time or tails all the time. It is the same way in genetics. The best statistical way that you can be sure to have nearly half of your children girls and the rest boys is to have many children, although some parents may reach this 1:1 ratio with as few as two children. There is still the distinct possibility, however, that if you have only two to four children, all will be boys or all girls. It is also possible for three of the children to be of one

sex and one of the other, or for two to be boys and two to be girls. Keep these facts in mind when you start to apply Mendelian genetics to human cases such as your own when you start to plan a family.

Determining Genotype and Phenotype of Progeny of a Mating by Means of a Punnet Square

An easy way to determine the relative proportions of genotypes and phenotypes to be found in the progeny resulting from a cross is to construct a Punnet square having as coordinates the genes from an allele set contributed by the gametes from both male and female parent organisms. Recall that in Meiosis I, the homologous pairs of doubling chromosomes proceed to separate; therefore, during this stage the genes of an allele set also separate. For example, consider the formation of gametes from an Rr individual. In Figure 13-2 you can see that in Meiosis I, the doubling chromosome containing the R gene separates from the doubling chromosome having the r gene of that allele set. Thus, from an individual which is Rr for a particular trait, half of the gametes formed would have the R gene of an allele set, whereas the remaining gametes formed would have the r gene of that allele set. Thus, by the Punnet square method, we would

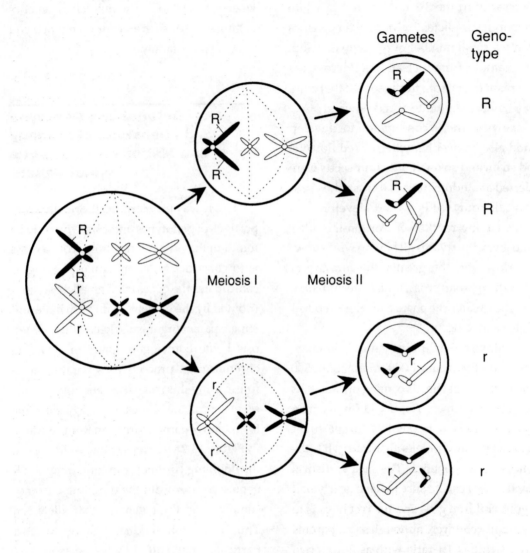

Figure 13 - 2 Separation of genes of an allele set Rr during meiosis to give gametes containing either of the genes R or r.

place the possible alleles of a cross between an Rr (female) X Rr (male) as follows:

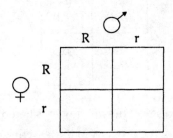

Now fill in the blanks of the square with the possible gene arrangement for each kind of progeny that can form as follows:

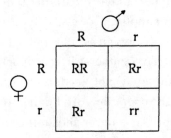

Hence, from this Punnet square, we can conclude that a cross of a plant heterozygous for red (Rr) with another Rr plant will give rise to progeny in which those with red flowers (RR and Rr) outnumber those with white flowers 3:1. The genotypic ratios according to the Punnet square is 1 RR : 2 Rr : 1 rr, or in other words, 1 homozygous dominant : 2 heterozygous : 1 homozygous recessive.

Once again I emphasize that the two letters RR, Rr or rr merely represent the two genes that are in an allele set and that are on a homologous pair of chromosomes in a diploid somatic cell or diploid fertilized egg.

The single letters in the coordinates of the Punnet square represent the one gene of the allele set that is in a gamete formed in meiosis and that is used in the cross. In fact, in modern terminology, *Mendel's first law* tells us that the genes of the allele set separate in Meiosis I. The Punnet square serves as a useful way to illustrate this process so long as you understand the procedures and symbols represented in the Punnet square.

Determining the Genotype of an Individual by a Test Cross

We can use the Punnet square to illustrate how it is possible to determine the genotype of an individual. In the case of the pea plant with white flowers there is no problem because by definition both genes in the allele set are homozygous recessive. We have a problem, however, when we consider pea plants with red flowers because they may be either homozygous dominant (RR) or heterozygous (Rr). But which one?

Here is how Mendel did it. He crossed plants with red flowers of an unknown genotype, that is either homozygous dominant (RR) or heterozygous (Rr), with plants having white flowers (rr). These are known to be homozygous recessive for the r gene. Then he analyzed the number of plants that had either red or white flowers among the progeny of the cross.

Let us analyze the two possibilities that Mendel encountered when he tried to determine the genotype of plants that may have been homozygous (RR) or heterozygous (Rr) red flowers. In either case, the plant with white flowers (homozygous recessive) contributes pollen that has only the recessive gene of the allele set we are considering; hence the top coordinates of the Punnet square consist of the letters r and r. The left hand coordinates, however, will be either R and R, or R and r depending on whether the red plant was homozygous dominant or heterozygous.

Hence if the red phenotype was due to the homozygous dominant genotype, the Punnet square of the test cross of the red RR with the white rr would appear as follows:

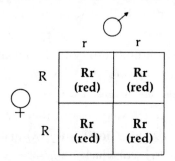

In other words, if the red flower is homozygous dominant, then all the progeny of the test cross will show the dominant phenotype.

On the other hand, if the plant having red flowers in the test cross was heterozygous, then the Punnet square would appear as follows:

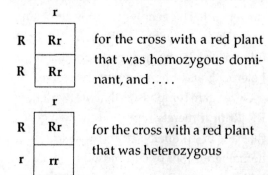

That is to say, if the red flower is heterozygous, half of the progeny of the test cross will show the dominant phenotype (red) and the other half will show the recessive phenotype (white).

Both of the above illustrations of *test crosses* are simplified for teaching purposes in that the phenotype as color of the flower is inserted in parentheses in each block; obviously, these are not needed in order to carry out the operations of a Punnet square. Furthermore, whenever illustrating a test cross by a Punnet square, the right hand column is redundant and can be eliminated because all of the gametes of a homozygous recessive organism have only the recessive gene of the allele being tested. Hence, the simplified Punnet squares would look like:

	r
R	Rr
R	Rr

for the cross with a red plant that was homozygous dominant, and . . . .

	r
R	Rr
r	rr

for the cross with a red plant that was heterozygous

Both give the same ratios as do the Punnet squares with two columns.

Finally, in practice, Mendel would test cross a plant of unknown genotype, but having red flowers, with a plant having white flowers. If all the progeny of the cross were red, then the genotype of the red plant used in the cross would be RR. If only half the progeny were red and the other half were white, then the genotype of the red plant used in the cross would be Rr.

and a test cross. Recall that the allele set represents the two genes, each on one of a pair of homologous chromosomes, which control the same trait; one of the genes in that allele set is on the chromosome which originally came from the female parent, and its allele is on the chromosome which originally came from the male parent.

Summary

You have just learned the most basic aspects of Mendelian genetics: how the two genes in an allele set affect the inheritance of the trait(s) controlled by a particular allele set in the progeny of a cross (mating). Of course Mendel did much more, for example, showing that the genes present in two different allele sets may *assort themselves independently* from each other so that the gametes from one individual are not necessarily the same with respect to the two genes of the different allele set. This principle, *Mendel's second law*, will be covered more fully when we discuss the genetics of human skin color in Chapter 15.

For now, however, make certain that you understand the terms genotype, phenotype, dominant, recessive, homozygous, and heterozygous, and how to do a Punnet square

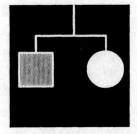

Chapter 14

Genetic Diseases; Sex Chromosomes

After learning the basic principles of Mendelian genetics, students usually try to apply those principles to the inheritance of human traits within their own family. To their disappointment, however, it soon becomes clear for a number of reasons that the study of human genetics is much more complicated than the study of the genetics of pea plants.

Some of the reasons are strictly practical. For example: (a) It is not usually possible to get large numbers of offspring from the matings of two individuals so that statistically significant numbers of children are obtained. (b) It is not practical, nor socially acceptable, to carry out a test cross. (c) The majority of observable or measurable human traits, such as intelligence, height, skin color, blood pressure, and finger prints, are *polygenic*, that is the result of the actions of many genes. Thus, few common traits can be classified as clearly dominant or recessive.

Nonetheless there are many interesting ways that we can study human genetics. First we will discuss a few common traits that are associated with *autosomal* (non-sex) chromosomes. Secondly we will discuss other traits that are also on autosomal chromosomes, but are expressed as *genetic diseases*. Lastly, we will discuss a number of instances in which a recessive gene on the X chromosomes is always expressed because the Y chromosome is devoid of the gene of that allele set.

Pedigree Analysis

Before studying human genetics, however, we have to overcome the handicap of not having sufficient progeny from two individuals to yield statistically significant Mendelian ratios. Hence for most studies in human genetics, we construct pedigree charts for as many generations as is possible, and then we try to analyze those family trees. An example of an extensive pedigree chart is shown in Figure 14-7, which illustrates the inheritance of hemophilia among the British, Spanish, and Russian royalty.

In pedigree charts for a human genetic disorder, a male is referred to by a square,   , a female by a circle,   , the affected (homozygous recessive) individual by a

darkened square, ■ or circle ●, and the sequence of generations by Roman numerals. An analysis of a pedigree chart will tell the relationship of individuals to each other; by showing the relationship of afflicted individuals to other members of the family, the pedigree chart may help in predicting the possibility that a descendant will be afflicted, or will be a carrier (heterozygous) of a particular gene. Some geneticists use half darkened squares, ◨ or circles, ◓, to indicate a carrier.

Autosomal Dominant Traits

Such traits require one gene of an allele set on autosomal (not X or Y) chromosomes to be expressed. In pea plants, the R gene controlling red color is an example of such a gene. In humans, however, most of these autosomal dominant traits can be detected only through biochemical studies. Most single genes which control autosomal dominant traits that are easy to observe in an individual are generally rare.

Yet there are some easily observable human traits that seem to be controlled by single autosomal genes: a few are listed in Table 14-1. The first column lists the dominant trait, and the second column lists the

Table 14-1	
Autosomal Dominant Traits and Complementary Recessive Traits	
Dominant	**Recessive**
Free earlobe	Attached earlobe
Freckled face	Clear face
Dimples	No dimple
Cleft in chin	Round chin
Widow's peak	Straight hairline
Curly hair	Straight hair
Dark eyes	Blue eyes
Middigital hairs	Absence of middigital hairs
Tongue rolling	Inability to roll tongue
Nonbending thumb	Hitchhiker's thumb

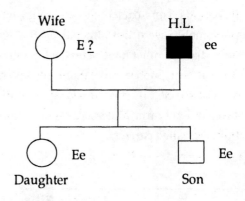

Figure 14 - 1 Pedigree chart for free and attached earlobes in author's family.

shows that my wife and two children have free earlobes; therefore, they all must have at least one dominant gene (E) in their allele set controlling free or attached earlobes. From the chart, therefore, is it possible to determine the genotype for that allele set for every individual in my family? Examination of the chart shows that it is possible to determine my genotype, that of my daughter and son, but not that of my wife. Do you know why?

trait usually expressed when the genes in the allele set are homozygous recessive.

Prepare a pedigree chart for one or more of these traits for your family. For example, I prepared one for the inheritance of free (Ee or EE) or attached (ee) earlobes in my family because I have attached earlobes, and, therefore, am homozygous recessive (ee) for this gene. Thus, all of my children received one recessive gene (e) of the allele set from me. The pedigree chart for my family (Fig. 14-1)

An Introduction to Genetic Diseases

For autosomal homozygous recessive genetic diseases to be expressed in a child, the fertilized egg must receive one recessive gene of that particular allele set from each of its parents. Thus, each of those parents, who presumably were not affected by that genetic disease, must have been heterozygous for that trait. Such people, that is those who are not affected by a particular genetic disease, and yet carry a recessive gene for that trait, are called *carriers.*

Cystic Fibrosis

As an example, consider the genetic disease called cystic fibrosis. Most individuals who are affected by this disease, that is who are homozygous recessive (cc) for that allele set, do not live to reach the age of 20. A six year old with cystic fibrosis may appear perfectly normal. Various symptoms like pneumonia, asthma, and malnutrition then start appearing. Other symptoms are high levels of salt in their sweat, obstruction of the intestines, pancreatic problems preventing digestion of food, clogged lungs, and heart strain. Life soon becomes extremely difficult for patients with cystic fibrosis. They require electric moisturizers to spray the room so that they can sleep. Their chests have to be clapped so that they can breathe. They may require as many as 20-40 pills a day in order to survive. They need calcium supplements and have extreme dental problems. Cough syrup is required to help keep their respiratory passageways clear.

In short, it may cost well over $50,000 a year to keep a patient with cystic fibrosis alive. There is much financial and emotional strain on the parents. Three out of every four parents having a child with cystic fibrosis end up divorced; some have nervous breakdowns because of the pressure. Others have great feelings of guilt because they know that they are carriers of the gene. Their misfortune is greatly compounded by the observation that children with cystic fibrosis usually have average or above average intelligence.

But how are parents to know that they carry the gene that might lead to some of their children having cystic fibrosis? To date there is no way for testing parents for the single recessive gene, for testing the cells obtained from an embryo by amniocentesis, or for detecting products of those genes. We know that one out of twenty-two people in the USA are carriers of the recessive gene (c) for cystic fibrosis. Thus, the chances are that in one in about 500 marriages (22 x 22=484) both parents will be heterozygous carriers (Cc) of the recessive gene. According to Mendel's laws and the following Punnet square,

Mother X	Father
Cc	Cc

	C	c
C	CC	Cc
c	Cc	cc

if the parents have four children, the odds are that one will not be affected and will not be a carrier (CC), two will not be affected, but will be carriers (Cc), and one will be affected with cystic fibrosis (cc). Hence, for an individual who is a carrier, the odds are 1 to 500 that he or she will marry another carrier, and then another 1 to 4 that one of the carriers' children will be affected. Thus, if you are a carrier, the odds are $1/500 \times 1/4 = 1$ to 2,000 that you will have an affected child.

Even if both parents are carriers, there is no guarantee that only one of four children will be homozygous recessive for cystic fibrosis. Remember, the Mendelian ratios only apply to populations; the larger the population the greater chance that the expected ratios will hold. If the carrier parents, on the other hand, have only one, two or even three children, it is possible that all these could be affected with cystic fibrosis, or that all three could become carriers, or that all three could become neither affected nor carriers. At the present time there is simply no way of knowing beforehand whether or not a person is a carrier, even if he or she is the child of two carriers. Hopefully, as the biotechnology of gene cloning advances, it may be possible to detect the recessive gene in carriers.

Measuring for Carriers of Genetic Diseases

There are 2,000 genetic diseases known today, and there are probably more. Of these we know of over 100 serious autosomal homozygous recessive diseases that are associated with the loss or alteration of a specific critical enzyme or protein. In a few of these diseases the heterozygote carrier exhibits a significant difference in the amount that he or she possesses of the critical enzyme involved. One such case is that of Tay-Sachs disease, a serious disease that will lead to the death of the affected child by age 2-4 through a series of degenerative processes of the nervous system. There is no cure for this disease.

We now know that the dominant gene leads to the production of an enzyme, which we will call "Tay-Sachs enzyme," that is critical for the formation of normal nerve cells. Only a single dose of that enzyme, a dose provided by one dominant gene, is needed for a person to be normal. Most individuals have a double dose of the enzyme because they have a double dose (TT) of the gene controlling formation of the enzyme, that is they are homozygous dominant for that gene. The carriers of Tay-Sachs disease, however, are heterozygotic for that gene (Tt) and therefore only produce a single dose of the Tay-Sachs enzyme. Thus, though normal themselves, they may transmit the recessive gene to their offspring. Knowing these facts, it is now possible to screen parents to determine whether or not they may be carriers of Tay-Sachs disease. A simple blood test will show that the blood of carriers contains about half the level of the Tay-Sachs enzyme than does the blood of non-carriers.

What if a newlywed husband and wife find out by means of a blood test that they both are carriers for Tay-Sachs disease, that they both are Tt. Does this mean that they must decide not to have children, or perhaps even divorce? What are the alternatives? Should they risk having children? After all, the odds are that 1 out of 4 of their children, even their first one, would have Tay-Sachs disease.

One answer to their dilemma lies with amniocentesis. Because there is a chemical test for the Tay-Sachs enzyme, it is possible to sample amniotic cells for the enzyme. If it is present in large amounts, the child will be normal for that trait. If it is present in intermediate amounts, the child will develop normally, but will be a carrier of the Tay-Sachs gene. If the enzyme is absent, then it can be predicted with great certainty that the child, when developed, will be affected with Tay-Sachs syndrome. In the latter case, the parents will then need to make a decision on whether to abort the embryo or not.

Frequency of Genetic Diseases

Our use of the Tay-Sachs disease as an example of an autosomal recessive trait that can be determined by a simple blood test, also helps point out the genetic advantage of marrying outside of the family. For example, since close relatives may be heterozygous for some of the same autosomal recessive traits, and thus be carriers, it is recommended the marriages between close relatives, such as first cousins, be avoided. The genes for the Tay-Sachs disorder are relatively more common among Jews whose ancestors once lived in Eastern Europe (the so-called Ashkenazi Jews) than in the rest of the population. It is estimated that 1 out of 30 Ashkenazi Jews are carriers of the recessive gene for Tay-Sachs disease, whereas in other groups only 1 out of 400 individuals are carriers. The frequency of genetic diseases is usually greater among members of inbred ethnic groups who intermarry frequently because of social situations and geography. For example, Ashkenazi Jews, who were usually isolated from their non-Jewish neighbors, are known to have a relatively high frequency of at least 19 distinctive genetic diseases. Blacks in general are known to have at least 7 common genetic diseases; African Blacks, 12; Eskimos, 4; Northern Europeans, 2; American Indians, 3; Chinese, 3; Japanese, 8; and Caucasians in general, 16.

Table 14 - 2	
Relationship	Proportion of genes in common
Identical twin	100%
Parent, child, brother, sister	50%
Fraternal twin	50%
Grandparent, grandchild, uncle, aunt, nephew, niece, halfbrother or half-sister	25%
First cousin	12.5%
Second cousin	3.1%

Hence, if a small community showed an increased incidence of any autosomal recessive genetic disease, the prudent thing would be for people not to marry close relatives, or better yet, to marry outside of that community. Table 14-2 shows the proportion of genes that various relatives have in common. From the table, it is obvious that so far as the prevention of autosomal recessive diseases is concerned, a marriage between second cousins is four times safer than a marriage between first cousins.

Will Chemical Tests for a Genetic Disease Help Eliminate that Gene from the Population?

On first thought one would think that if parents knew that they were carriers of a genetic disease, then they would decide not to have children, and consequently, the frequency of that gene in the population would decrease.

To the contrary. First consider what usually happened to two married carriers of Tay-Sachs disease before the test for detecting the gene in carriers was discovered. Assume that their first child was healthy, and, therefore, was either free of the Tay-Sachs gene (TT), or was a carrier(Tt). Next, assume that the second child was affected by Tay-Sachs disease (tt). Now those parents would have 2 to 4 very difficult years until the affected child died. Furthermore, they could be burdened by guilt and worry that should they have another child, that it too would be affected by Tay-Sachs disease. Hence, the parents most likely would decide to have no more children of their own. As a result, they would introduce no more Tay-Sachs recessive genes into the population.

Today, however, the situation is very different. Carrier parents can test by amniocentesis to determine whether or not the prospective child would be affected with Tay-Sachs disease and decide whether to abort or not. Consequently, with this means of determining the genetics of their prospective children and with the guarantee that none of their children will be affected with the disease, those carrier parents may decide to have an average size family of 2-3 children. If we assume that large numbers of such carrier parents decide to have 3 children, then Mendelian genetics tells us that the odds are that 2 of every 3 of those children will be carriers (Tt) of the recessive genes. Therefore, we can see that with good tests available for a recessive gene which can cause a genetic disease when homozygous, those genes may actually increase in the general population rather than decrease.

Incomplete Dominance of an Autosomal Recessive Genetic Disease

Recall that each gene leads to the production of one protein. In the case of autosomal recessive diseases like Tay-Sachs, one dose of the "good" protein produced by the dominant gene is sufficient to make that individual as healthy as a person who is homozygous dominant and who thus has a double dose of the good protein. There are a number of other genetic diseases, however, in which the heterozygous condition of one dominant and one recessive gene will lead to a condition of *incomplete dominance*. In

some of these cases, the homozygous dominant individual will be healthy, the homozygous recessive individual will be affected by the genetic disease, and the heterozygous individual will be somewhere in between. One of the better known examples is the genetic disease of *sickle cell anemia* (see Chapter 16).

Other instances of intermediate inheritance can be found in Mendel's studies. For example, he crossed one type of a homozygous dominant red plant with a homozygous recessive white one and found all the progeny to be pink. In this case, let us call the dominant gene S for scarlet, with ss being white. Thus, in the progeny of this cross there was incomplete dominance because the protein produced by one S gene was not sufficient to give the plant a deep red color, and instead a pink color (Ss) was obtained. In the case of this plant, it takes a double dose of the dominant gene, as in the homozygous dominant condition (SS), for that plant to produce enough protein for it to have a deep red color.

Traits Linked to the X Chromosome

Up until now we have been discussing traits associated with the autosomal chromosomes. In those cases both genes in an allele set were important in determining whether or not a trait would be expressed. For example, in the autosomal dominant traits, at least one of the two genes in the allele set had to be dominant. In the autosomal recessive traits, both genes in the allele set had to be recessive for the trait to be expressed; an individual heterozygous for that trait is considered a carrier.

Although the situation with X-linked traits in the case of females is similar to that of autosomal recessive traits, with males the situation is different. Recall that females have two X chromosomes, but that males have only one X chromosome and one Y chromosome. Thus, for every gene on an X chromosome in females, there exists the other gene of the allele set on the female's other X-chromosome. In males, however, the Y chromosome is very small compared to the X chromosome. Thus, for virtually all of the genes on the X chromosome in the cell of a male, the Y chromosome does not have the second gene required for an allele set.

What is the consequence of this situation? Consider the case of a recessive gene on the X chromosome that caused green-red colorblindness. For this trait to be expressed in a female, both X chromosomes would need to contain the recessive gene of the allele set. Thus, a homozygous recessive female, which we will indicate by $X^c X^c$, would be colorblind. A heterozygous female, $X^C X^c$,

would not be colorblind but would be a carrier just as she would be if the gene were carried on an autosomal chromosome.

In males, however, an individual $X^c Y$ having only one recessive gene of the allele set for colorblindness, would be colorblind. Why? Because the Y chromosome has no gene, either dominant or recessive, in the allele set, to affect the phenotype with regards to colorblindness. Thus, in order for a male not to be colorblind, he must receive an X chromosome that has the dominant gene; that is, he must be $X^C Y$. It is not surprising to learn, therefore, that whereas in the United States only 1% of females are colorblind, over 8% of males are.

X-linked Traits in Progeny from Parents of Differing Genotype:

Example (a), described in Figure 14-2, tells us of the possibilities that would arise in the genotype and phenotype of children from a father normal for the X-linked trait and a mother who is a carrier for that trait. In this case, there are equal chances for having a normal female, a carrier female, a normal male, or an affected male. Four other situations can arise:

(b) With a father normal for the trait and a mother who is affected, then all the sons will

Figure 14 - 2

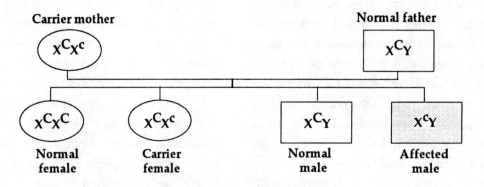

Figure 14 - 3

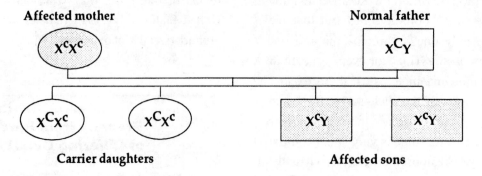

Figure 14 - 4

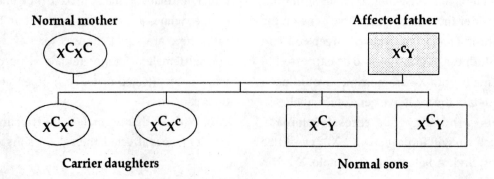

Figure 14 - 5

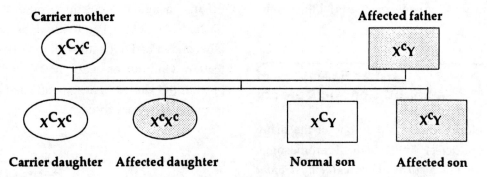

Figure 14 - 6

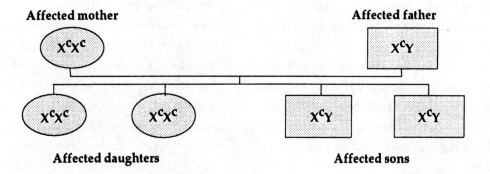

be affected and all the daughters will be carriers (Fig. 14-3).

(c) When an affected father and a mother normal for the trait, all the sons will be normal, and all the daughters will be carriers (Fig. 14-4).

(d) A most complex situation occurs with an affected father and a mother who is a carrier. In such a case, half of their sons will be normal, half of their sons will be affected, half of their daughters will be affected, and half of their daughters will be carriers (Fig. 14-5)

(e) Finally, with an affected father and an affected mother, obviously all of their children will be affected (Fig. 14-6).

Thus, from these pedigree charts, it is apparent that with regards to X-linked genetic diseases, the male is at a distinct disadvantage. About the only advantage to being male in this regard is in the case in which one son is affected and the other is not; then the unaffected son can be certain that he is not a carrier. The unaffected daughters of such a family, on the other hand, will not know whether or not they are carriers until they have sons of their own, or, in some cases, from accurate pedigree charts for the family.

Other Examples of X-linked Genetic Defects

Colorblindness is only one of the many genetic defects linked to the X chromosome. Another is a form of *muscular dystrophy* called Duchene's muscular dystrophy. Symptoms are first observed in the baby who appears clumsy. By age 6, the muscles of the children affected with Duchene's muscular dystrophy lack firm muscles; between the ages 9-12, they become restricted to wheelchairs; and by age 20 the majority die, although some live past the age of 50. There is no known treatment for this disease that affects about 3 males out of every 10,000 births.

Another serious X-linked trait is *immune deficiency disease.* This disease has gained great publicity because of a young man known as "David." Immediately on his birth in 1971, he was placed in isolation from all other living creatures in a protective germ-free "bubble" in a Texas hospital. He remained there for 13 years, when he was finally let out of this protected environment for an operation and finally to have human contact, such as his first kiss from his mother. He died shortly after the operation. David,

like others born with immune deficiency disease, was not able to make antibodies to fight off any bacteria or virus that would invade his body. Hence, such individuals are virtually condemned to death with their first infection. One out of every 10,000 children born in the United States are affected with immune deficiency disease.

Hemophilia

The blood of individuals affected with this X-linked disease usually takes 5 to 10 times longer to clot than does the blood of non-hemophiliacs. Thus, a cut, an extracted tooth or a bloody nose can lead to massive hemorrhaging which could cause death. That hemophilia was transmitted in certain families among males was noted many centuries ago in the Hebrew Talmud; the Talmud prescribed that if a male bled excessively after being circumcised, then subsequent boys born into that family were exempt from circumcision.

We now know that the recessive gene on the X-chromosome leading to hemophilia is unable to make an important protein in the blood needed for clotting, whereas the dominant gene does make that important protein. In past times the majority of hemophiliacs never reached adulthood. Today, however, most reach adulthood because they can be injected with a factor isolated from normal

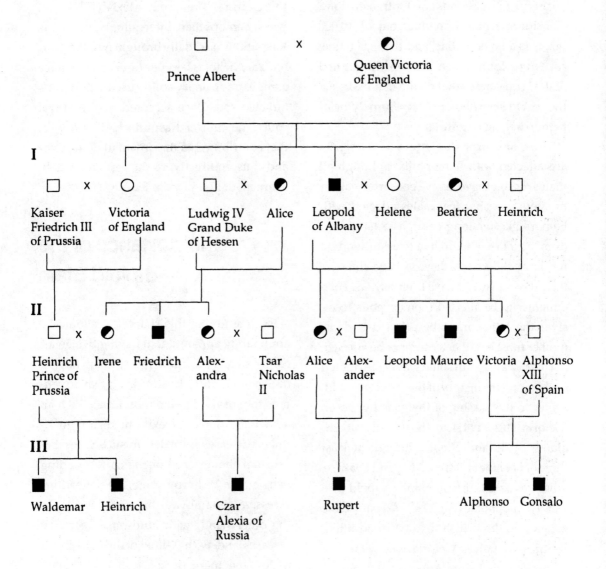

Figure 14 - 7

blood that will hasten clotting. Some hemophiliacs may have to spend over $40,000 a year for such injections and for transfusions. Considering one individual out of 10,000 births is a hemophiliac, and that 8,000 new cases are found each year in the United States, you can see that on the national scale the cost to keep hemophiliacs from dying of hemorrhaging is quite large.

There are surprisingly few females who are affected with hemophilia, although all males receive the recessive gene from carrier mothers. Possibly few males affected with hemophilia live long enough to have children, or those who do live to adulthood do not have children. If they do have children, their wives may not have been carriers. For a daughter to be affected (homozygous recessive) the father must be affected and the mother must be either a carrier or be affected.

This disease has in some ways contributed to changing the history of the modern world. I refer to the passing of the gene by Queen Victoria (1819-1901) to the royalty of England, Spain and Russia through at least three of her nine children: Alice and Beatrice, who proved to be carriers, and Leopold who was a hemophiliac (Fig. 14-7). All passed the gene on to the British royalty. In addition, Beatrice's daughter Victoria, who was also a carrier, married King Alfonso XII of Spain. She transmitted the gene to their son, Prince Alfonso, who died at the age of 19 after an automobile accident.

Alice's daughter Alexandra married Nicholas II, the last Czar of Russia and became the Czarina. Their son, Alexis (1904-1916) was a hemophiliac. Interestingly, the monk Rasputin was initially brought into the court of Czar Nicholas because he claimed to have some extraordinary ability to cure the Czar's afflicted son. Once accepted into the royal court, the sinister Rasputin had an impact that contributed to the downfall of the Czar and thus, indirectly, to the resultant establishment of Leninism in Russia.

Importance of the Diploid State

From studying this chapter, it should be abundantly apparent that having two genes controlling a function rather than having one gene do the job, affords a great advantage for survival to the individual. Such an advantage does not exist in regard to X-linked diseases in males; in such cases, because of the absence of the corresponding gene on the Y chromosome, the deleterious recessive gene always expresses itself. On the other hand, with autosomal recessive diseases and with X-linked diseases in females, the mere presence of one "good" dominant gene in the allele set guarantees in most cases that the deleterious gene will not express itself in that individual. Thus, by getting one complete set of genes from the 23 chromosomes donated by one parent, and another complete set of genes from the other parent, an individual has a fifty-fifty chance that he or she will not be affected by a deleterious gene passed on by either of the parents. If, on the other hand, humans were haploid instead of diploid, that is if humans had only one set of chromosomes in their cells, then their chances of expressing recessive genes that lead to genetic diseases would be very great. Hence, as pointed out in Chapter 1, one of the functions of sex is to promote survival of the species. The other function of sex, that of enhancing genetic variability, will become much clearer in the next chapter.

Behavioral Genetics

There has been much research recently showing a genetic link with certain types of behavior. One case was mentioned on page 141 in which a microdeletion, i.e. a removal of one small portion of a chromosome, was found to be present in one chromosome (chromosome 7) of all individuals tested who have Williams syndrome. People with Williams syndrome are extremely friendly and outgoing, have a great facility with languages, and have varying degrees of talent for music. On the other hand, they can not do simple mathematics, and lack many so-

cial skills. None of their normal parents tested showed that microdeletion.

Another case concerns homosexuality. There seems to be growing evidence of a genetic link for the behaviors associated with homosexuality. Much of the evidence comes from the study of twins. In identical twins (see p. 233), if one has homosexual tendencies, then there is a good chance that the other twin will show the same behaviors, even if reared by another family. The same does not hold true for non-identical twins in which one may be homosexual. Other research indicates that genes located on the X chromosome may be linked to sexual orientation. Still other research shows that a tiny region of the hypothalamus in the brain may be smaller in homosexual men than in heterosexual men.

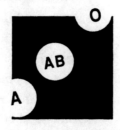

CHAPTER 15

Traits Controlled by More than One Gene

The study of human genetics is much more complex than are the examples of single allele sets that we have been studying thus far in Chapters 13 and 14. Most traits that we can observe and measure in humans, such as stature, I.Q., and height, are affected by many genes; that is, they are *polygenic*. In order to begin to introduce you to the excitement and complexities of understanding genetics in humans, I will present in this chapter five examples dealing with but two genes, that is, two allele sets.

(a) **Example I**: Two different allele sets on two different pairs of homologous chromosomes controlling *one trait*.

(b) **Example II**: Two different allele sets on two different pairs of homologous chromosomes controlling *two different traits*.

(c) **Example III**: *Two different allele sets* which are situated *close to each other* on each chromosome of a pair of homologous chromosomes, and which control two different traits.

(d) Example IV: *Two different allele sets*, each controlling a different trait, which are situated *far apart from each other* on each chromosome of a pair of homologous chromosomes.

(e) Example V: *One allele set* on one pair of homologous chromosomes, which at any one time will be made up of any two of *three possible allelic genes for that trait.*

To make these cases a little more meaningful and interesting to study, I will use as examples allele sets that control for such traits as skin color, the easily observable traits of attached or free earlobes, the genetic disease known as galactosemia, and the production of the A, B, AB and O blood types.

Example I: Two Different Allele Sets on Two Different Pairs of Homologous Chromosomes Controlling One Trait

For our first example, consider a simplified but illustrative case of the genes controlling the pigmentation of human skin. Although scientists believe there are at least four allele sets which control the production of the pigments that give the color of our skin, in our example we will pretend, for the

sake of simplification, that only two allele sets, AA and BB, are involved.

In this simplified example. let us define a person whose skin is *black* as homozygous dominant for both of these allele sets, that is, for black skin the genotype will be AABB. At the other end of the spectrum, we define a person with *white* skin as homozygous recessive for both allele sets, i.e. aabb. Skin colors in between black and white will depend upon the relative number of the dominant and recessive genes. Thus, we define a person with *dark* skin as AABb or AaBB (i.e. 3 dominant: 1 recessive); *intermediate* skin color as AaBb, AAbb, or aaBB (i.e. 2 dominant:2 recessive); and *light* skin color as Aabb or aaBb (i.e. 1 dominant:3 recessive).

Consider the children of a woman with white skin (aabb) and a man with black skin (AABB) (Fig. 15-1). For the two genes affect-

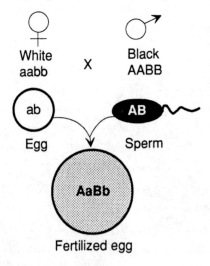

Figure 15 - 1 Genotype of zygote of white female and black male.

Table 15-1 **Possible Genotypes and Phenotypes Among the Progeny**				

eggs \ sperm	**AB**	**Ab**	**aB**	**ab**
AB	AABB Black	AABb Dark	AaBB Dark	AaBb Intermediate
Ab	AABb Dark	AAbb Intermediate	AaBb Intermediate	Aabb Light
aB	AaBB Dark	AaBb Intermediate	aaBB Intermediate	aaBb Light
ab	AaBb Intermediate	Aabb Light	aaBb Light	aabb White

ing skin pigmentation, the genotype of the woman's eggs would be ab and that of the man's sperm would be AB . Their children would all possess skin of intermediate pigmentation (AaBb).

This example differs from those dealing with single allele sets as in Chapters 13 and 14. In this case *each of the allele sets are on different pairs of homologous chromosomes* (Fig.15-2). Each parent produces gametes of the same genotype as far as these two chromosomes are concerned.

Now let us consider a genetically more complex case, one involving two individuals of intermediate skin color (AaBb). The question we will ask is "What proportion of children from such parents will be black, dark, intermediate, light, or white in skin color?"

First we must distinguish the genotype of the eggs and sperm that will be involved in such a mating. From the simplified diagram in Figure 15-3, it is obvious that each parent can produce gametes of four genotypes, as far as the two genes for skin color are concerned: AB, Ab, aB and ab. If you feel more comfortable drawing each chromosome, you may do so, but I think that you will find the method used in (Fig. 15-3) easier to use.

Next let us construct a Punnet square just as we did in Chapter 13, but this time we need to consider gametes of four genotypes (Table 15-1) rather than two.

Analysis of the Punnet square (Table 15-1) shows that rather than getting a 3:1 ratio of genotypes as we would get when crossing two Rr plants, in this case we get a more

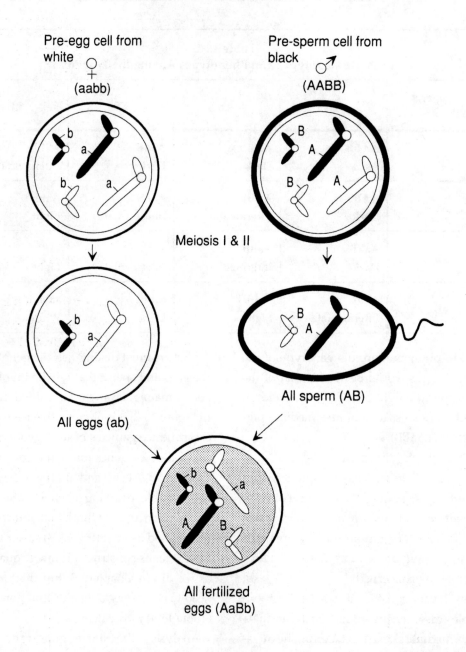

Figure 15 - 2 Chromosomes and genotype of gametes and zygote of white female and black male.

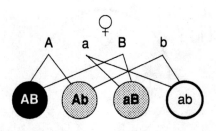

Figure 15 - 3 Genotypes of gametes of female with intermediate skin color (AaBb).

complex ratio of phenotypes in the offspring: 1 black: 4 dark: 6 intermediate : 4 light: 1 white (Table 15-2).

Example II: Two Different Allele Sets on Two Different Pairs of Homologous Chromosomes Controlling Two Different Traits

Now let us consider an example in which each of the two allele sets controls a different trait, rather than one trait as was just described with skin color. We will now use a hypothetical case in which one allele set, LL or Ll, controls for *detached ear lobes*, and the homozygous recessive allele set, ll, controls for *attached ear lobes*.

The other allele set of genes in this hypothetical example is the one controlling the formation of the enzyme which allows chil-

Table 15-2 Proportion of Possible Phenotypes Among the Progeny					
Ratio of dominant to recessive:	4:0	3:1	2:2	1:3	0:4
Genotypes:	AABB	AABb	AaBb	Aabb	aabb
		AABb	AaBb	Aabb	
		AaBB	AaBb	aaBb	
		AaBB	AaBb	aaBb	
			AAbb		
			aaBB		
Proportion of Phenotypes:	1 Black:	4 Dark:	6 Intermediate:	4 light:	1 white

dren to digest the galactose sugars found in milk. Children who are homozygous dominant (GG) or heterozygous (Gg) can digest the sugar, whereas children who are homozygous recessive (gg) cannot. These latter children are affected with the genetic disease known as *galactosemia*. As a consequence of their inability to digest galactose, which is a sugar found naturally in milk from humans and cows, the brain of these children will not develop properly and they will become mentally retarded unless they are placed on a diet free of galactose during their early years of childhood.

In our example, each allele set of genes is on a different set of homologous chromosomes, as shown in Figures 15-4 and 15-5. We are interested in finding the distribution of these two traits in children of couples who are heterozygous for both of these allele sets. To find these distributions we go through exactly the same steps as we did for skin color in Figure 15-3 and Table 15-2. In the present example, however, since the genes of the two allele sets do not control the expression of the same trait, the Punnet square that we construct (Table 15-3) will give different ratios of the various possible phenotypes.

This phenotypic ratio of 9:3:3:1 (Table 15-4) is typical for a cross of plants, or matings of animals, in which we follow the distribution of two traits which are controlled by genes of two different allele sets found on two different pairs of homologous chromosomes.

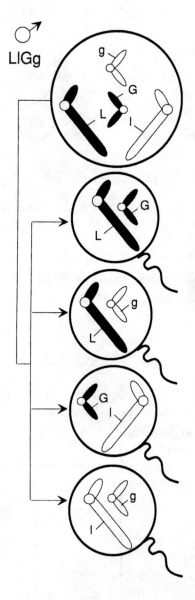

Figure 15 - 4 Distribution of two pairs of homologous chromosomes during Meiosis I and II leading to the formation of sperm. The black chromosomes signify that they were of maternal origin, whereas the colorless ones signify that they were of paternal origin.

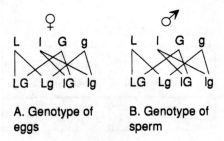

A. Genotype of eggs

B. Genotype of sperm

Figure 15 - 5 Genotype of eggs (A) and sperm (B) from parents heterozygous for free ear lobes and absence of galactosemia.

Mendel, using pea plants, first noticed this type of phenotypic distribution of two traits. He concluded that the two sets of factors, that is genes of two allele sets, involved in such a cross *assorted* themselves in the gametes independently of each other. In other words, he essentially predicted that in meiosis there is no way of knowing in advance whether a gamete receives a maternally or paternally contributed chromosome (see Fig. 15-4).

This phenomenon of *independent assortment* of maternal or paternal chromosomes with their respective genes into gametes, is often referred to as *Mendel's second law*. We now can say, therefore, that Mendel's second law refers to the random manner in which chromosomes of maternal and paternal origin assort themselves into gametes during Meiosis I. *Mendel's first law*, if you recall, refers to the separation of the two genes of an allele set also during Meiosis I.

Example III: Two Different Allele Sets Situated Close to each other on One Pair of Homologous Chromosomes

When two genes which control different traits, one from each of two different allele sets, are situated very close to each other on

Table 15-3 Possible progeny of children developing from eggs and sperm described in Figure 15-5				
Eggs \ Sperm	LG	Lg	lG	lg
LG	LLGG	LLGg	LlGG	LlGg
Lg	LLGg	LLgg	LlGg	Llgg
lG	LlGG	LlGg	llGG	llGg
lg	LlGg	Llgg	llGg	llgg

Table 15-4
Ratio of Phenotypes of Children from Crosses Described in Table 15-3

	Free lobe, no galactosemia	Free lobe, galactosemia	Attached lobe, no galactosemia	Attached lobe, galactosemia
	LLGG	LLgg	llGG	llgg
	LLGg	Llgg	llGg	
	LLGg	Llgg	llGg	
	LlGG			
	LlGG			
	LlGg			
	LlGg			
	LlGg			
	LlGg			
Ratio of Phenotypes:	9	3	3	1

the same chromosome, they almost always stay with each other when the chromosomes assort themselves during meiosis. This sort of behavior is in direct contrast to that shown in Example II where the two different allele sets were located on two different pairs of homologous chromosomes; in that previous example those chromosomes randomly assorted themselves during meiosis.

Furthermore, we can say that those two genes of those two allele sets which are situ-ated close to one another on the same chromosome are *tightly linked*. Let us use the same two allele sets of genes described in the last example, the one causing free or attached ear lobes and the one that could lead to the genetic disease of galactosemia. In this example the mother and father are both heterozygous for both of these traits, and I have specifically selected a situation in which one of the homologous chromosomes of the pair (the one originating maternally) has both

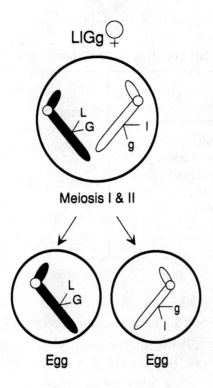

Meiosis I & II

Egg   Egg

Figure 15 - 6 The transfer of genes of two allele sets (Ll and Gg) on a pair of homologous chromosomes during meiosis.

dominant genes, L and G, and the other homologous chromosome of that pair (the one originating paternally) has both recessive genes, l and g.

From Figure 15-6 it is clear that an individual having a genotype of LlGg has the two dominant genes, one from each allele set, on one of the homologous chromosomes, and the two recessive genes, one from each allele set, on the other homologous chromosome. It is also clear from the figure that L and G stay together on the same chromosome dur-

ing meiosis, and that l and g do the same on the other homologous chromosome. Thus, Figure 15-7 shows the genotypes of the two kinds of gametes that each LlGg parent can form.

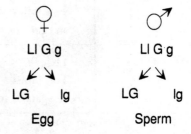

Figure 15 - 7 Genotypes of gametes produced by individuals who are LlGg with those genes situated as in Fig. 15-6.

The Punnet square (Table 15-5) shows that the distribution of the two traits which occur in the progeny follow a 3:1 ratio. This is the same Mendelian ratio found in a heterozygous mating in which the genes of one allele set for one trait were followed as described in Chapter 13. The difference is that in the present example we are following two traits which are controlled by the genes of two separate allele sets. The 3:1 distributions show that the genes controlling those traits are *tightly linked*, that is, the genes of the two different allele sets are situated close to each other on the same pair of homologous chromosomes.

Table 15-5
Distribution in Progeny of Two Traits Tightly Linked on Same Chromosome

Eggs \ Sperm	LG	lg	Final ratio:
LG	LLGG Free lobes; not galactosemic	LlGg Free lobes; not galactosemic	3 with free lobes; not galactosemic to 1 with attached lobes; galactosemic
lg	LlGg Free lobes; not galactosemic	llgg Attached lobes; galactosemic	

Linkage

It may appear to you that I am belaboring a rather straightforward point. I do this to give you an appreciation of the practical advantages of understanding the genetic phenomenon of linkage. As you will see, linkage may help physicians identify a newborn child who has a life threatening genetic disease in sufficient time to save the life of that child.

Let us use the hypothetical example given in Table 5-5 of the genes controlling for free (LL or Ll) or attached (ll) ear lobes, and for the presence (gg) or absence (GG or Gg) of galactosemia. Suppose for the family under consideration that physicians knew that the recessive form of those two genes (l and g) were tightly linked on the same chromosome, and that the dominant form of those two genes (L and G) were closely linked on

the other homologous chromosome. Thus, if the trait for the homozygous recessive genotype (ll for attached ear lobes) was expressed, then they could be certain that the other allele set of tightly linked genes was also present in the homozygous recessive state (gg for galactosemia).

Thus, if parents with free ear lobes (Ll) have a child with attached ear lobes (ll), then the physicians could be certain that the child also would be homozygous recessive for galactosemia (gg) and unable to digest the sugar galactose. Such children will become retarded if they are given a diet of either mother's or cow's milk. In this case, then, the physician would prescribe a diet free of galactose, as in soy bean "milk" and other foods, until the child was of an age that its brain was fully developed.

Although this hypothetical example to show how knowledge of linkage could prevent a child from becoming retarded was concocted strictly for teaching purposes, the concept is a real one. Physicians use a battery of chemical tests to determine linkages for genes controlling the expression of certain proteins when the genetic histories of the parents warrant such tests.

The study of linkage is also valuable to help us better understand: (a) the order of genes on a chromosome; (b) other mechanisms for providing greater genetic variation for the survival of the species; and (c) another important complexity of meiosis, that of crossing over (see below). These three aspects of linkage can be seen more clearly in Example IV in which the genes of the two allele sets in question are situated relatively far apart on the same pair of homologous chromosomes.

Case IV: Genes of Two Different Allele Sets Situated Far Apart from each Other on each Chromosome of One Pair of Homologous Chromosomes

For this hypothetical example, we will consider genes of the same two allele sets that we described in Cases II and III, that is, the genes affecting free (LL or Ll) or attached (ll) ear lobes, and the absence (GG or Gg) or presence (gg) of galactosemia. In this Case IV, however, the genes of those allele sets

will be located far apart on each chromosome of the homologous pair (Fig. 15-11B).

A. Tightly linked genes

B. Loosely linked genes

Figure 15 - 8 Locations of tightly (A) and loosely (B) linked genes on pair of homologous chromosomes.

What will be the consequences of such an arrangement compared to that in which those same two genes were situated very close to each other, i.e. the situation in which they were "tightly linked" (Fig. 15-8A)? Will the genes that are far apart also be linked tightly and thus stay together on the same chromosomes of the homologous pair as they segregate during the formation of gametes? The answer is yes.... sometimes, and no . . . sometimes. If this answer does not seem satisfactory, remember that genetics follows the laws of probability, and that if you study enough cases, you will find that some sort of pattern will emerge.

In the case of two different genes on the same chromosome, the rules are: (1) The closer that two genes (of two allele sets) controlling different functions are situated near one another on the same chromosome of a homologous pair, the greater the chance that they will remain situated that way by the end of meiosis. (2) Conversely, the further apart those two genes are situated from one another on the same chromosome, the greater the chances that they will not remain on the same chromosome of a homologous pair by the end of meiosis, i.e., that they are not tightly linked.

Just what does it mean when we say that those two genes will not remain on the same chromosome of a homologous pair by the end of meiosis? What happens to them? Does one disappear? No. Does the chromosome break in half? Not exactly. Then what does happen? The answer is that during the first meiotic division, maternal and paternal portions of the pair of homologous doubling-chromosomes which house those genes, intertwine and exchange equal pieces. To get a better understanding of this complex phenomenon, look at Figure 15-9A-E. Let us carefully follow the steps that occur in the separation of the pieces of the intertwined chromatids from the homologous doubling-chromosome containing genes G,

g, L and l. You can see that the exchange took place at the point at which one of the chromatids of a doubling-chromosome of maternal origin (black) crossed over and contacted one of the chromatids of a doubling-chromosome of paternal origin (colorless).

The major result of the exchange was that the four gametes formed from the one pre-gamete cell which exhibited the crossing over shown in Figure 15-9D, all had a different genotype as regards to the genes G, g, L and l. What does this mean as far as you are concerned? It means that through the process of crossing over, you produce gametes (egg or sperm) having a greater variety of genetic information than you would have had if no crossing over had occurred. This genetic variation, if you recall from Chapter 1, is the major reason for sexual reproduction, i.e. genetic variation increases the chances that the species will survive. It also contributes greatly to the fact that your children are not exactly alike genetically.

Possibilities for Crossing Over

Figure 15-9 illustrates only one of the possibilities for crossing over in one pair of homologous doubling-chromosomes. Not only is it possible for the chromatid arms of most 23 pairs of homologous doubling-chromosome to cross over each other during mei-

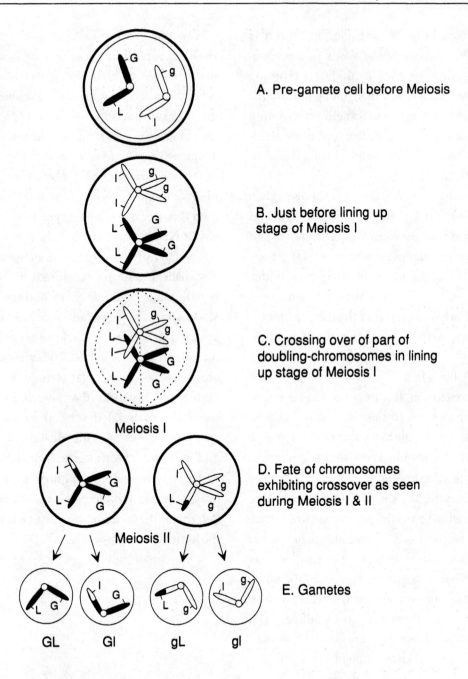

A. Pre-gamete cell before Meiosis

B. Just before lining up stage of Meiosis I

C. Crossing over of part of doubling-chromosomes in lining up stage of Meiosis I

Meiosis I

D. Fate of chromosomes exhibiting crossover as seen during Meiosis I & II

Meiosis II

E. Gametes

GL   Gl   gL   gl

Figure 15 - 9 Crossing over of parts of chromatids of homologous chromosomes, and its effect on the distribution in the gametes of loosely linked genes.

osis, especially the long ones, but it is also possible for one homologous pair of doubling-chromosomes to undergo crossing over at virtually any place along the lengths of their chromatids. Sometimes crossing over can occur at more than one place along the arms of the chromatids during the same meiotic division.

Crossing over can occur: (1) at more than one point on doubling-chromosomes; (2) on either side of the centromere; (3) more than once on one side of a centromere; (4) without disturbing the order of the genes under study; (5) hardly ever between two genes which are very close to each other (i.e. tightly linked); (6) frequently between genes (allele sets) that are relatively far apart (i.e. not tightly linked).

In conclusion, it is important to be aware of the concept of crossing over as it occurs in meiosis during the formation of the eggs and sperm. You should know that it is a major factor leading to genetic variability, which is necessary for the survival and evolution of the species. Secondly, you should know that through the study of crossing over, biologists are able to estimate the relative distances between different genes on the same chromosome - the greater the occurrence of crossover between two genes (allele sets), the further apart they are on the chromosome. Thirdly, knowledge of linkage is important in some medical diagnoses.

Now that you feel that you have grasped the basic aspects of crossing over, have you ever thought of how biologists determine whether or not crossing over occurs between genes of two particular allele sets? No, they do not see the genes on the chromosomes. Think again. The answer is that they carry out genetic crosses just as Mendel did, they observe the phenotype of the progeny, and then they deduce the genotype of the progeny. A typical human couple, however, does not usually produce sufficient offspring for a statistical analysis of crossover. Hence, with humans, pedigree charts and statistical studies on larger populations of individuals are used to determine the location of genes on human chromosomes. Today biologists are using more molecular techniques, techniques which require that you first understand the material described in the next chapter, to determine the location of genes on the human chromosome. The U. S. government is now contemplating a gigantic project to allow biologists to identify and determine the location of every gene on all the human chromosomes.

Case V: One Allele Set that at any One Time will be Made up of any Two of Three Possible Allelic Genes for a Specific Trait

Such a situation is found in the allele set controlling the blood types A, B, AB and O. First let us make certain that you understand the concept of blood type and the consequences of transfusing blood of the wrong type into an individual's blood stream.

Our red blood cells are classified by the type of proteins found on their surface (Fig. 15-10). The red blood cells of some individuals possess on their surfaces a protein called A protein; their blood is called type A. Blood having red cells with the B protein on its surfaces is type B; with both proteins, type AB; and with neither of these proteins, type O.

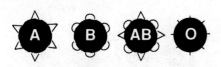

Figure 15 - 10 Schematic representation of red blood cells with antigens A, B, AB or none on their surfaces.

Antibodies for these proteins normally exist in the blood of humans; these antibodies do not need to be induced by outside agents as do most other antibodies. Recall that *antibodies* are proteins that are formed in individuals in response to foreign proteins, or material, such as viruses or bacteria,

which get into our bloodstream. Antibodies in general exist as free proteins dissolved in the fluids of our blood, i. e. *serum*.

In addition to antibodies formed in response to foreign materials, human blood serum contains a number of antibodies which occur naturally. Among the naturally occurring antibodies are the antibodies to the A and B proteins of red blood cells (Fig. 15-11). The serum of a person with type A blood has only anti-B antibodies (Fig. 15-11). Accordingly, type B blood has only anti-A antibodies; type AB blood has neither anti-A nor anti-B antibodies; and type O blood has both anti-A and anti-B antibodies (Fig. 15-12 and Table 15-6, columns 1, 2, 3).

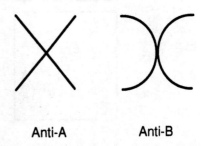

Anti-A   Anti-B

Figure 15 - 11 Schematic representation of anti-A and anti-B antibodies.

A major problem occurs if blood of the wrong type is given to an individual by transfusion. For example (follow Table 15-6, first row), if a person of blood type A (column 1) receives blood that is either of type B or type AB (column 4), the anti-B antibodies

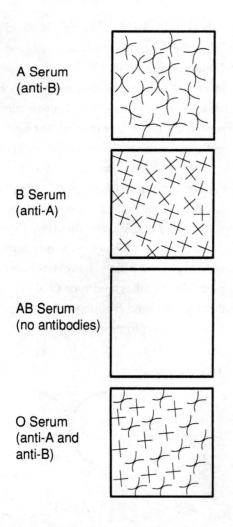

A Serum
(anti-B)

B Serum
(anti-A)

AB Serum
(no antibodies)

O Serum
(anti-A and
anti-B)

Figure 15 - 12 Serum from type A, B, AB, and O individuals showing antibodies.

In either case, these clumped cells might then lodge in small blood vessels of the recipient and cause serious consequences often leading to death. On the other hand, column (5) shows that type A blood can be received by persons of either type A or type AB because the blood of type A or type AB individuals does not have anti-A in their fluids. [Note: The little bit of anti-B antibod-

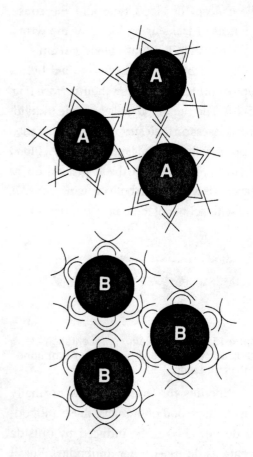

Figure 15 - 13 Clumping of type A red blood cells by anti-A antibodies, and type B cells by anti-B antibodies.

present in type A's blood (column 3) will react with the B or AB transfused cells and will cause them to clump together (Fig. 15-13). A similar clumping of type A red blood cells would occur if type A blood were introduced into the blood of a type B individual.

Table 15-6					
Type of Blood	**Complete Blood**				
(1)	**(2)** Protein on red blood cells	**(3)** Antibody in blood fluids	**(4)** Blood types that will cause clumping if donated to blood type of column (1).	**(5)** Blood types of individuals who will be safe recipients for blood type of column (1).	**(6)** Blood type that can be donated to individuals of blood type of column (1).
A	A	Anti-B	B and AB	A, AB	A, O
B	B	Anti-A	A and AB	B, AB	B, O
AB	A and B	Neither anti-A nor anti-B	None	AB	AB, A, B, O
O	Neither A nor B	Anti-A and anti-B	A, B and AB	O, AB, A, B	O

ies that would enter the blood of a type AB person receiving type A blood would be of a too low concentration to affect the AB blood cells of recipient type AB.]

As another example, the last column of Table 15-6 shows that a person with type O blood, that is, a person whose red blood cells have neither A or B protein on their surface (column 2, last row) will be accepted by individuals of types A, B, AB or O. Why? Because type O red blood cells have no proteins on their surface that can be recognized and clumped by anti-A or anti-B antibodies. In short, people with type O can give blood to individuals of all types (column 5), and can be called *universal donors*. On the other hand, type O individuals are very restricted in the type of blood they can receive. A person with an AB blood type, on the other hand (row 3), does not produce either anti-A or anti-B (column 4), and, therefore, is a *universal recipient* since the blood contains neither anti-A nor anti-B antibodies and consequently does not clump transfused blood of types A, B, AB or O.

It is difficult to understand the biology of transfusions unless you think in terms of "relative proportions" and *"dilutions."* For

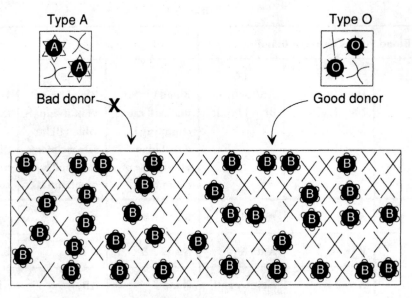

Type A

Type O

Bad donor — X

Good donor

Recipient's blood - Type B

Figure 15 - 14 Example of how donor's antibodies are diluted in recipient whereas donor's red blood cells may be clumped by recipient's antibodies.

example, the antibodies in one pint of donated blood will become so diluted in the recipient, that they will have no effect on the recipient's red blood cells. Consider the case illustrated in Figure 15-14 in which a recipient of blood type B will receive either a small amount of blood of type A or of type O. When type A blood is given to a type B individual, the type A red blood cells will be tied up by the anti-A antibodies in the blood of the type B recipient and will clump; the type B antibodies that go into the bloodstream of the type B recipient, however, are in such a small concentration that they have no significant effect on the receipient's type B red blood cells.

If, on the other hand, type O blood is given to the type B individual, the type O red blood cells will not be clumped by the anti-A antibodies in the recipients serum; also, the small amount of anti-A and anti-B antibodies that came along in the donor's type O blood are in such a low concentration that they will have no effect on the type B red blood cells of the recipient. If, however, the recipient was of type O blood and was given a transfusion of type B blood, the donated red blood cells would clump. Do you see why?

Genetics of Blood Types A, B, AB and O

Geneticists have learned that the allele set controlling the type of protein on the surface of the red blood cell consists of any two of three genes, not two as we have been studying thus far. In this situation, which we refer to as *multiple alleles*, only two of the three genes, however, are present on a pair of homologous chromosomes in an individual at any one time. The A gene will lead to the formation of the A protein, the B gene will lead to the formation of the B protein, and the O gene will lead to the formation of an O protein.

For an individual to be of blood type A, he or she must have a genotype homozygous for A, i.e. AA, or heterozygous for A and the inactive O, i.e. AO. Likewise, the genotype for a person who is of blood type B must possess either BB or BO genotypes. It stands to reason that the genotype of a person who is of blood type AB is AB. Such a person is heterozygous for the A and B genes; since both genes of the allele set express themselves, we speak of the AB genotypes as exhibiting codominance. Finally, the geno-

type of a person with O blood type must be OO.

As an example, consider the possible blood types of progeny of a marriage between a woman of blood type O with a man of blood type A. Since we do not know the genotype of the man, which can be either AA or AO, we can construct two Punnet squares.

The first Punnet square will tell us that if all the children have type A blood type, then the father's genotype is most likely AA. On the other hand, the second Punnet square will tell us that if some of the children have type A blood and others have type O blood, then the genotype of the father is AO.

With this kind of information readily available by a simple blood test, you can see why such blood tests are often used in paternity and rape cases. For example, what if a woman with type O blood was raped and had a child as a result of the rape who also had type O blood. A year later a suspect, who was a married man, was accused of the rape. Examination of the blood of the accused father and his children showed conclusively that the father was genotypically AA. Our Punnet square will tell us that a man with an AA genotype could not have been the father of a child with blood type O. Thus, the suspect was exonerated of the crime.

There are other examples of how knowledge of blood types can be used to settle legal cases. For example, could a male with type O blood rightfully claim to be the father of a

child who has an A blood type? a blood type
B? a blood type AB? The answers are yes,
yes, and no. Can you figure out why?

CHAPTER 16

Molecular Genetics

Today much of genetics can be explained at the molecular level. By understanding some of this chemistry, you will have a better understanding of how genes work, of the nature of mutations, and of the molecular basis of genetic diseases. You have already had an introduction to molecular genetics in the last part of Chapter 9. I urge you to study that material and understand it before reading further. Then you will be prepared to delve into the underlying molecular mechanisms of genetics.

Basic Terminology and Assumptions

DNA (deoxyribonucleic acid), a long double stranded molecule found in chromosomes, is the material that makes up the gene. DNA functions to make specific proteins via two types of single stranded molecules called RNA (ribonucleic acid). That is, every gene, which is a segment of a double-stranded DNA molecule, directs the synthesis of a single strand of a specific RNA, called *messenger RNA* (mRNA). The messenger RNA, with the help of another RNA, called *transfer RNA* (tRNA), arranges an array of single amino acids to line up in a specific order. Next, these amino acids are linked together to form a chain of *amino acids* called a *protein*. This protein chain then coils up into its own specific shape. Most of these proteins are *enzymes* (see Chapt. 9).

Molecular geneticists would summarize the above processes by stating that DNA, the gene, *transcribes* its information into RNA which then *translates* that information to arrange amino acids to form a specific protein.

The DNA also contains the information to make more of itself, that is to replicate. We see this during the preparatory stage of either mitosis or meiosis, when the DNA molecules in the chromosomes *replicate* to make exact copies of themselves. In this way they form the *doubling-chromosomes* seen in cells undergoing mitosis or meiosis.

Origin of Some Concepts of Molecular Biology

The first suggestion that genes control the formation of proteins came at the turn of the century from a number of studies of disorders of human heredity. One of these early works, for example, dealt with the observation that children who could not metabolize one common amino acid became mentally retarded.

The discoverer of this correlation, a Dr. A. Garrod, called such a disorder an *inborn error of metabolism*. He postulated that genes control the formation of enzymes, and that if a gene mutated and was expressed in a child, then that child would suffer from one of those inborn errors.

We now know that most inborn errors are genetic diseases caused by the expression of a recessive gene. We have already discussed one such genetic disease in Chapter 15. That disease, *galactosemia*, is an inherited disease in which an infant lacks the enzyme to metabolize the sugar galactose, which comes from milk.

Once physicians learned that galactosemia is caused by the lack of one enzyme, they devised a simple blood test for that enzyme. Currently newborn infants are routinely tested for the presence of the enzyme. If the test shows a newborn to lack that enzyme, the infant is placed on a diet free of milk and galactose. If the afflicted child is not placed on that restricted diet, it will become mentally retarded, and develop cataracts and a defective liver.

Sickle Cell Anemia

The study of another genetic disease, *sickle cell anemia*, led to discoveries which have greatly advanced our knowledge of how the gene works. Sickle cell anemia is a recessive autosomal disease that affects mostly blacks of both sexes. In some communities in Africa, 40% of the population is affected, in the USA, about 8%. The red blood cells of afflicted individuals, rather than having the normal shape of a disk, at conditions of low oxygen take on the quarter-moon shape of a sickle. Individuals who are homozygous recessive for the trait are at great risk. Those who are heterozygous, show the effects of the disease only under extremely stressful conditions.

When an afflicted individual is exposed to conditions of low oxygen, the red blood cells take on the sickle-shape, clog the small capillaries and then a cascade of symptoms occurs. As a consequence, during a medical crisis, which is often fatal, a patient develops a fever and intense pain in the bones, stomach, and joints. Frequently heart disease and kidney failure result. Some of the many effects of such crises are shown in Figure 16-1.

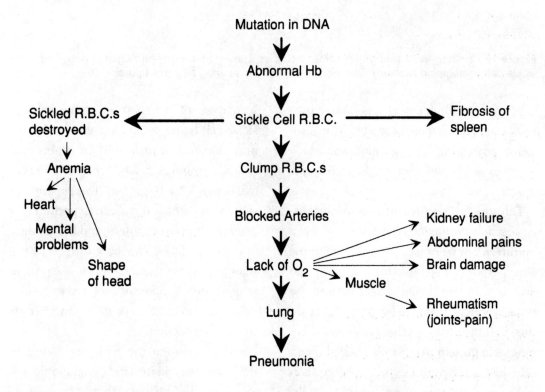

Figure 16 - 1 Many of the pleiotropic effects generated by the single mutation causing sickle cell anemia.

Before mutation:

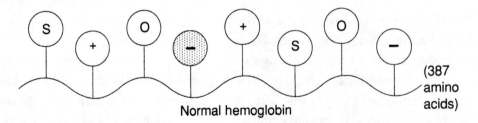

Normal hemoglobin

(387 amino acids)

After mutation:

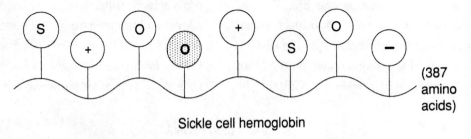

Sickle cell hemoglobin

(387 amino acids)

Figure 16 - 2 Charges of side groups of amino acids making up normal hemoglobin (top) and sickle cell hemoglobin (bottom). See Figure 9-6 for help in analyzing this figure.

We refer to the many effects caused by a mutation, such as those occurring in an individual possessing the single mutation which leads to sickle cell anemia, as *pleiotropic effects.*

Through further investigation of this disease, scientists obtained the first proof that a *mutation* can be caused by a change in the gene , a change that leads to only one amino acid in a protein being different from the protein normally formed by that gene. To do this, first they showed that the purified hemoglobin protein from sickle shaped blood cells had a different electrical charge than did hemoglobin from normal red blood cells.

Analysis of the chain of amino acids in sickle cell hemoglobin showed it to have a non-charged amino acid in place of a charged amino acid which is found in normal hemoglobin (Fig. 16-2). This experiment showed that a gene dictates the specific *order of amino acids* in a protein, and that a mutation in the DNA can be expressed by a change in just one amino acid in the protein in question. And because of a change in a single amino acid, a protein can alter its shape and function.

Reflecting upon the high prevalence of sickle cell anemia in Africa, one should ask why a deleterious gene should be so common. The answer lies with the unusual properties given the red blood cells of individuals who carry even one of the sickle cell genes. Apparently the malaria parasite, which has to go through one of its life stages inside a red blood cell, does not do well in red blood cells possessing the sickle cell hemoglobin. Because malaria is a major killer of humans in Africa, individuals with the sickle cell trait would have a greater chance of surviving longer than would individuals with normal red blood cells. Hence, the gene is preserved, and increases in frequency in African populations.

Chemical Structure of DNA

Much evidence has accumulated that the gene is DNA. This double-stranded molecule is composed of two chains of a *polymer* of four different *nucleotides*. A polymer is defined as a long molecule that consists of many units of a smaller molecules of similar structure. For example, starch, which is made by plants, is a polymer consisting of a chain of the sugar molecule known as glucose. The polymer chains in DNA, in contrast, consist of alternating *sugar* and *phosphate* molecules. In its simplest form, a representation of this polymer is shown in Fig. 16-3A. In it, the sugar molecule, which is the five carbon sugar *deoxyribose*, is represented by a vertical line, and the phos-

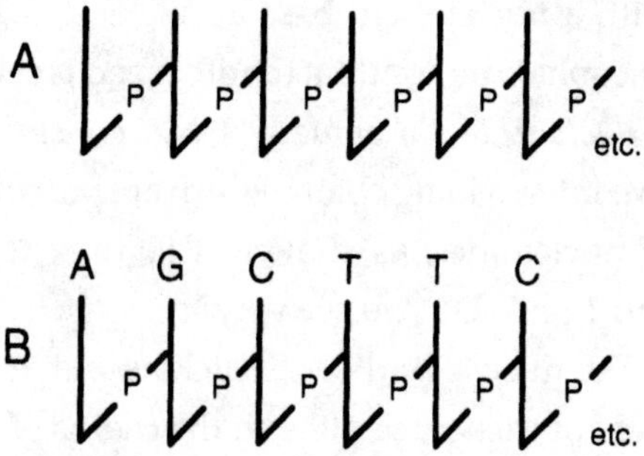

Figure 16 - 3 Schematic representation of polymer of sugar phosphates (A) and nucleotides (B).

phate, which connects every other sugar molecule, is represented by the letter P.

At one end of each sugar molecule in the polymer is attached a *nucleic acid base*, labeled in Figure 16-3B as either A, G, C or T. Those letters represent the molecules a̲denine, g̲uanine, c̲ytosine and t̲hymine. *Adenine* and *guanine* are double ring-like structures called *purines*, whereas *cytosine* and *thymine* are single ring-like structures called *pyrimidines*.

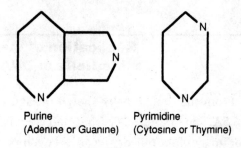

Purine
(Adenine or Guanine)

Pyrimidine
(Cytosine or Thymine)

Figure 16 - 4 Schematic representation of a purine and a pyrimidine.

The single unit of a deoxyribose sugar with a nucleic acid base at one end, and a phosphate molecule at the other end is called a *nucleotide*. So actually DNA consists of two intertwining chains which are polymers of nucleotides, as shown in Figures 16-3B and 16-5A. Do you see why?

The nucleic acid bases sticking out at the ends of each nucleotide in the chains of the DNA possess certain important physical and chemical properties. These cause the two opposing polymers of the double stranded molecule to wind around each other to form a *double helix* (Fig. 16-5A). From this figure you can see that the nucleic acid bases of the two strands oppose each other with an *A opposite a T, a G opposite a C, a T opposite an A, and a C opposite a G.* Actually those pairs of nucleic acid bases are attracted to each other by weak chemical forces. These forces, called *hydrogen bonds*, are indicated in the figure by dotted lines between each opposing pair of nucleic acid bases.

Replication of the Double Helix of DNA

From the double helix shown in Figure 16-5A, you can see that each strand is a mirror image of the other as regards the composition of the complementary nucleic acid bases. It is this *mirror-image* structure that is the foundation of the process by which DNA replicates itself. This replication must occur during the preparatory stage of mitosis and meiosis when the single chromosome becomes a doubling-chromosome.

During this phase, the double helix of DNA unravels at one end, and each strand replicates itself to form two double helixes (Fig. 16-5B). Here is how it happens (Fig. 16-5A). As the double helix unravels at one end, nucleotides floating free in the cell, move to the unraveled ends of the double helix and line up alongside the complementary nucleic acid bases of the two strands. That is, the A nucleotides are attracted to the T nucleic acid base of the unraveled strand, the G to the C, etc. By a special enzymatic process, those lined up nucleotides are zipped together, and they start to wind around the unraveled ends of the DNA to form two new double helixes of DNA (Fig. 16-5B). As this figure also shows, each of the two new double helixes that are formed during this process of replication, consist of one strand of the original DNA and another strand of the newly synthesized DNA strand, which is a mirror image of the original strand of DNA.

Remember, the process of replicating DNA takes place in every one of your forty-six chromosomes during mitosis and meiosis. Figure 16-6A illustrates replication of the DNA in a chromosome as it becomes a doubling-chromosome, and Figure 16-6B shows

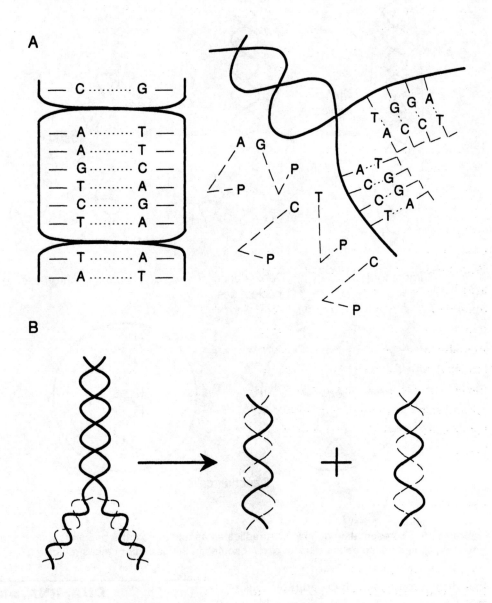

Figure 16 - 5 Replication of DNA showing the unraveling of the double helix (A and B), and how free nucleotides are incorporated into new strands of DNA (A), and how each new double helix of DNA retains one of the old strands (dark lines) and gains a new one (lighter lines) (B).

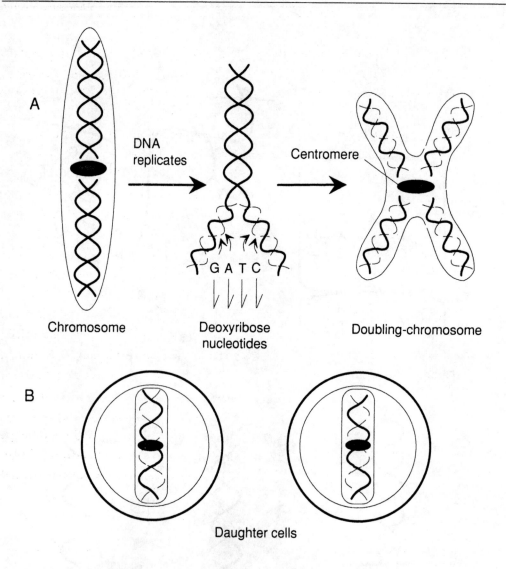

A

DNA replicates

Centromere

Chromosome

Deoxyribose nucleotides

G A T C

Doubling-chromosome

B

Daughter cells

Figure 16 - 6 Representation of how DNA replicates as a chromosome becomes a doubling chromosome (A), and becomes the DNA of chromosomes in the daughter cells of mitosis (B).

that chromosome in the two daughter cells. The DNA is drawn to indicate that it received one strand from the original double helix, and the other strand from the synthesis beginning with the free nucleotides.

DNA, RNA, and the Synthesis of Proteins

Recall (Chapt. 9) that a protein is a long chain of amino acids linked together in a very specific order; depending upon the order and properties of those amino acids, the long chain usually coils up into a ball-like structure which gives the protein its specific function. The key to forming proteins then, is to provide a mechanism for lining up amino acids in the proper order so that they can be linked into a chain. DNA and RNA provide that mechanism.

Before considering the steps by which a protein is made, you must first understand that a double helix of DNA consists of many genes. Each *gene* is essentially one relatively small portion of one of the two strands of the DNA double helix. For one gene to effect the synthesis of one protein, that gene must first transcribe the information within it to a free floating smaller molecule. The information within the gene resides in the order of its nucleotides, and the free floating molecule to which it transcribes that information is called *messenger RNA* (mRNA).

Messenger RNA is one three types of RNA molecules involved in the synthesis of a protein as initiated by a gene. The RNA stands for *ribonucleic acid*. RNA, like DNA, is also made up of nucleotides, but it differs from DNA in three general ways: (1) the sugar in RNA is *ribose*, which is also a five carbon sugar, but which differs from deoxyribose; (2) RNA does not utilize thymine for a nucleic acid base as does DNA, but uses another pyrimidine, *uracil*, instead; and, (3) RNA exists as a *single strand*, not as a double helix in the way DNA does.

Messenger RNA is formed in a manner similar to the way DNA replicates itself. First the strands of the double helix DNA unravel, but only in the region which represents the gene involved in making a protein (Fig. 16-7). Next free *ribose nucleotides*, not deoxyribose ones, line up at the region of the gene on only one of the unraveled strands. Instead of thymine nucleotides to match up with the

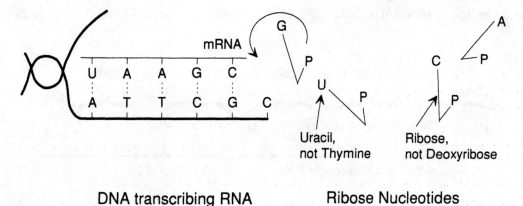

DNA transcribing RNA   Ribose Nucleotides

Figure 16 - 7 Transcription of DNA to form messenger RNA from ribose nucleotides.

adenines sticking out of the unraveled DNA strand as is the case in the replication of DNA, there will be *uracil* nucleotides. The figure shows the ribose nucleotides lining up along the nucleic acid bases of the gene on the unraveled strand of DNA to form a molecule of mRNA. The other types of RNA, that is, transfer RNA and ribosomal RNA described below, are formed by essentially the same process, which is called *transcription*. In transcription, a nucleotide with an A matches up with a U, a T with an A, a C with a G and G with a C.

Thus, we can say that the portion of DNA representing a gene serves as a *template* for making a single strand of mRNA. That strand is called messenger RNA because it has the "message" - i.e. the information to direct the synthesis of a protein - the end product of the gene. In fact, we can say that for every protein, there is a specific mRNA, and for every specific mRNA, there is a gene.

The message of mRNA is *translated* into a specific protein by means of the *genetic code*.

That code refers to groups of three nucleotides in the mRNA called *codons* (Fig. 16-8). As illustrated in this figure, each codon has a different order of the nucleotides, as reflected by their nucleic acid bases, such as UAA, GCG, etc.

Each codon serves to direct a specific amino acid to line up along the messenger RNA. For a codon to do this, however, requires the involvement of another type of RNA called *transfer RNA* (tRNA). Each of these RNAs is also made of a single strand, but the strand folds up in such a way that at one end there is place for an amino acid specific for that tRNA to attach itself; at the other end is a group of three nucleic acid bases called an *anti-codon* (Fig. 16-9). There are a little over twenty tRNAs, one with its respective anti-codon for each kind of amino acid used in making a protein.

Thus, once a gene is transcribed to make a specific mRNA, the codons on that mRNA attract the anti-codons of the various tRNAs; in these cases, the A, G, C, and U of the codon

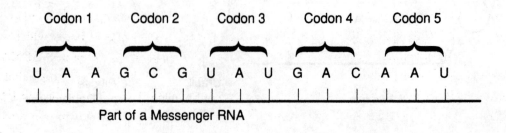

Codon 1   Codon 2   Codon 3   Codon 4   Codon 5

U A A G C G U A U G A C A A U

Part of a Messenger RNA

Figure 16 - 8 Five codons of three exposed nucleic acid bases, each along a chain of nucleotides in a messenger RNA.

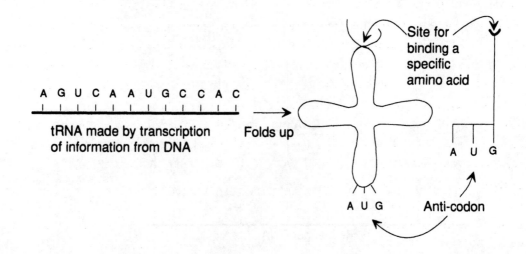

A G U C A A U G C C A C

tRNA made by transcription
of information from DNA

Folds up

Site for
binding a
specific
amino acid

A U G

A U G   Anti-codon

Figure 16 - 9 Schematic representation of transfer RNA showing anti-codon and site for binding amino acid.

match up with the U, C, G, and A of the anti-codon respectively. Each of those tRNAs which are attracted to the codon of the mRNA have a specific amino acid attached at their other end (Fig. 16-10).

Finally, the amino acids at the ends of those tRNAs attracted by their anti-codons to the complementary codons of the mRNA and lined up there, are bonded together to form a chain of amino acids by means of a round structure in the cell called a *ribosome* (Fig. 16-11A). The ribosome is made up both of RNA molecules and protein molecules that function in the last stages of protein synthesis. The new chain of amino acids which is bonded together by the ribosome, is released into the cell where it coils up because of the attractive forces on the amino acids. There it forms the protein originally

coded for by the gene that started this whole process (Fig. 16-11B).

Thus you have it: DNA transcribes its information to make RNA (mRNA) which translates that information, with the help of tRNA and ribosomes, to make protein!

Mutation

A *mutation* is defined as a change in one of the DNA nucleic acid bases which will lead to a permanent change in the DNA of that cell and all daughter cells once that altered DNA replicates. The replicated DNA with its permanent change in one nucleic acid base will direct the production of a different codon on the mRNA for that gene. As a consequence of that change in a single

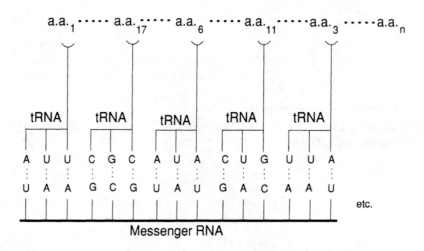

Figure 16 - 10 Lining up of five transfer RNAs along a messenger RNA. Note that the nucleic acid bases of the codons of mRNA and the anti-codons of tRNA attract each other. In the meantime, the amino acids carried at the opposite end of the tRNAs are lined up in order to form a protein. The dotted lines signify weak but significant chemical attractions.

nucleic acid base in the gene, the mRNA will differ by one nucleic acid base in just one of its codons. That new codon, in turn, will attract an anti-codon from a different tRNA with a different amino acid attached to it. Thus, the protein formed from that mutated gene will have, except for one amino acid, exactly the same amino acids as would have been made if that gene had not mutated. Nonetheless, the change in one amino acid can alter significantly the way that a newly formed protein coils to take its shape. As is often the case, the change of one amino acid will alter significantly the shape, and therefore the function, of the protein. A good example of such a mutation caused by the coding for a different amino acid can be seen as in the case of the gene coding for the formation of hemoglobin in individuals with sickle cell anemia (Fig. 16-2).

Mutations can occur spontaneously, more frequently with some genes than others. A study of 9 inherited diseases show that 4 spontaneous mutations occur per 100,000 gametes produced. Mutations that occur in gametes can be passed on from generation to generation if those gametes lead to the formation of a child. In addition, mutations can also occur in our somatic cells, some causing cancer. Mutations in somatic cells, except for pre-gamete cells, however, are not passed on to the next generation.

Mutations can be induced, i.e. their rate of formation can be speeded up, by a number

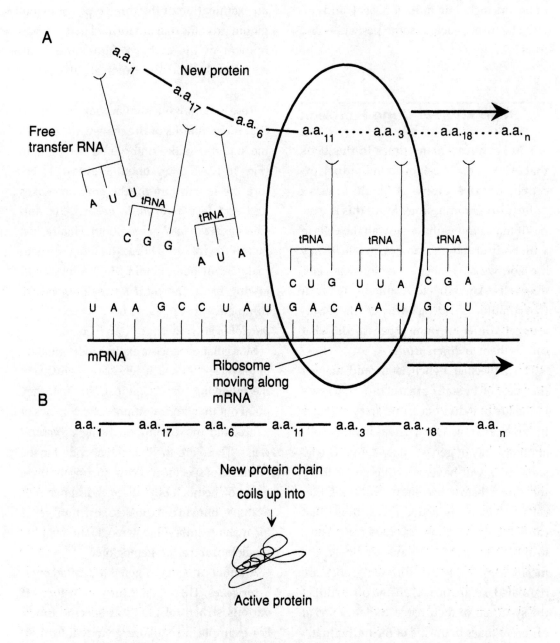

Figure 16 - 11 (A) Synthesis of a chain of amino acids by means of mRNA, tRNA, and ribosome, and (B) coiling of that chain to make a protein. See text. The bold short lines in between the amino acids (a.a.) signify strong chemical bonds, whereas the dotted lines signify weak attractions.

of factors including radiation, heat, and certain chemicals, such as some pesticides and food additives.

Regulation of Gene Function

You have read many times in this book that every somatic cell in an individual, except in the cases of mosaics, has the same set of forty-six chromosomes. Since this is true, has it made you wonder why all the cells in your body are not the same? Obviously they are not; we have nerve cells, muscle cells, liver cells, skin cells, etc. Another way to put that conundrum is to ask why are those various cells different, or, in other words, what causes them to *differentiate*?

The molecular geneticist would answer the question by stating that at different times in the life of individuals, one array of genes in some cells is active, whereas in other cells another array of genes is active. In this way some cells will have as active genes those that direct them to become nerve cells. Other cells will have as active genes those that cause them to become red blood cells. Thus, in molecular terms, the answer lies in the mechanisms by which different genes are regulated, i.e. are turned on and off, to direct the synthesis of their respective proteins at different times in the life of the individual.

Good examples of these molecular mechanisms for controlling genes can be found in an examination of the three types of hemoglobin proteins that are formed in the development of the embryo, fetus, and adult respectively, and the times at which they appear.

Researchers have shown that during the first three months of pregnancy, the cells in the embryo make embryonic hemoglobin (Fig. 16-12A). Later, about a few weeks before the fourth month, the embryo makes less and less *embryonic hemoglobin* and starts to make fetal hemoglobin. Finally, just before the infant is born, the fetus starts to make adult hemoglobin. By 3 months following birth, the *fetal hemoglobin* nearly completely disappears and only *adult hemoglobin* is synthesized from then on.

Molecular geneticists explain the regulation of the means by which genes are expressed in the following way (Fig. 16-12B). First, they point out that there a number of categories of genes, the most common being *structural genes*. These are the genes that code for the synthesis of specific proteins, such as the three types of hemoglobins discussed above. For example, human cells possess structural genes for making embryonic hemoglobin, fetal hemoglobin, and adult hemoglobin.

Another category of genes are called *regulator genes*. These genes turn on or turn off various structural genes at specific times. For example, a regulator gene will turn on the structural gene controlling the formation of embryonic hemoglobin only during the

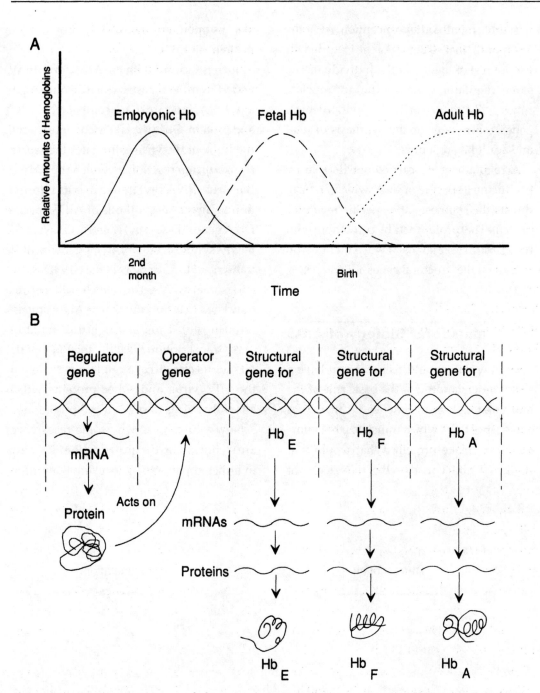

Figure 16 - 12 Genetic regulation of synthesis of three different hemoglobins (see text).

first three months after conception. After the 3rd month, that structural gene is turned off for the rest of the life of the individual. The same regulator gene, at the appropriate times, will also turn on and off the structural genes controlling for the synthesis of fetal and adult hemoglobins.

A regulator gene carries out its role by producing a specific mRNA, which, in turn, directs the synthesis of a specific *regulator protein*. That protein acts by combining with the *operator gene* so that it will either turn on or turn off the structural genes adjoining it.

Molecular Biology of AIDS

By now you should feel that you have a good understanding of the basic rule of the molecular biology of genetics, i.e. that DNA transcribes RNA which translates that information to make protein. What would you think if I told you that the rule does not

always operate in the direction of DNA to protein via RNA?

Such is the case with the AIDS virus (HIV) and a few other viruses which do not contain any DNA, but rather are composed of RNA and protein. Once the HIV virus enters a cell, the RNA of the virus stimulates the *reverse transcription* reaction by which the RNA of the virus (RNA HIV) transcribes its information to direct the synthesis of AIDS-specific DNA genes (DNA HIV) (Fig. 16-13). Next the newly synthesized DNA HIV genes will be transcribed to form more of the HIV RNA and and some mRNA HIV which will become translated to form a number of AIDS proteins (Protein HIV). Thus, through this initial reverse transcription reaction, the RNA of the HIV virus can direct a cell to make more of the HIV virus and other proteins which allow that virus to survive and proliferate.

Now you can see why some researchers are trying to stop the growth of the HIV virus in afflicted patients by focusing their atten-

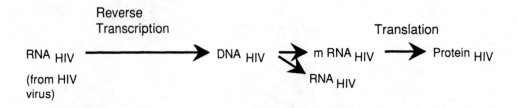

Figure 16 - 13 Reverse transcription by AIDS virus (HIV).

tion on designing and manufacturing drugs which will stop the enzyme controlling the reverse transcription reaction.

Biotechnology

Molecular genetics is now going through some exciting times. Scientists are attempting to apply the newly acquired information from this field toward improving our living standards and health. For example, it is possible today to identify specific genes in chromosomes, to remove those genes from the cell, to identify the chemical structure of those genes, and to synthesize them in the laboratory.

It is also possible to insert some of those isolated genes into the chromosomes of other organisms. The potential for these new technologies seems endless. For example, by inserting one kind of gene into a plant, it may be possible to make a plant that is resistant to certain pests and diseases, or to make a plant which will provide an abundant and more nutritious food source.

Another application is to isolate a gene which directs the synthesis of an important hormone, such as insulin, or a cancer-fighting protein, and to insert that gene into an easy-to-grow harmless bacteria. The goal would be for the bacteria to make those life-saving proteins inexpensively so that they could be available to all who need them.

Finally, experiments are under way to insert genes for making specific proteins into humans who, because of a genetic disease, may lack those genes. Much more research and testing must be done. None of those technologies are as simple as one might hope. Further, there are many legal and ethical questions that must be answered before we enter fully into the age of the new biotechnology.

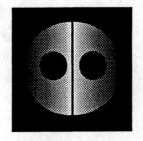

Chapter 17

Fertilization, Early Development, and Twinning

Fertilization

Fertilization can be defined simply as a two-step process consisting of: (1) the union of the haploid sperm nucleus with the haploid nucleus of the egg to form the diploid nucleus of the *zygote*, and (2) the initiation of the development of the embryo. Although the events of fertilization are much more complex than implied in that brief definition, it does convey the major point that fertilization consists of two important events, a genetic one and a metabolic and cellular one.

The genetic event is the recurring theme that we have been emphasizing throughout this book: the fusion of the genetic material of maternal and paternal origin to provide the resulting embryo with the continuity of the parents' genetic lines, and the *genetic variability* resulting from the mixing of the two parental gene pools.

The other important event in fertilization is the *activation* of the cellular and metabolic (biochemical) processes which allow the development of the embryo from the fertilized egg to proceed. The activation process can even be made to occur in some eggs without the nucleus of the sperm entering the egg. For example, biologists have activated unfertilized eggs of a number of species by "shocking" them with certain chemicals, or even by a pin prick; those activated eggs would then go on to divide. The natural phenomenon by which eggs of certain organisms, such as aphids, spontaneously start to divide and develop a viable organism without the involvement of the sperm is called *parthenogenesis*, a form of asexual reproduction that involves only the egg (see p. 3).

There is a rare form of parthenogenetic activation occurring in a fish called the Amazon Molly, however, that does involve an unusual use of sperm. These fish are all females; no males of the species exist. In order to activate her eggs, the Amazon Molly mates with males of a closely related species of fish. The nucleus of their sperm enters the egg, thereby activating it, but does not fuse with the nucleus of the egg. Instead, the sperm nucleus eventually disintegrates, the egg (really the secondary oocyte) does not get rid of its polar body; the secondary oocyte and polar body become one cell, and their haploid nucleuses fuse to give one diploid nucleus. At that time the resultant "fertilized" egg starts to divide to form the diploid embryo, and, eventually, a mature female Amazon Molly.

Events of Fertilization in Humans

Both the egg and the sperm must be ready for fertilization. For example, the *membrane* surrounding a mature egg (secondary oocyte) must be able to fuse with the membrane of the sperm when the egg and sperm come in contact. In addition, the *cytoplasm* of the mature secondary oocyte must be able to accept the sperm nucleus, support the fusion of the sperm and egg nucleuses, and allow the chromosomes to duplicate and the events of mitosis to take place.

The mature sperm, on the other hand, cannot fertilize an egg until it is given that capacity by residing about seven hours (the residency time) in the fluids of the female genital tract. Biologists believe that through this *capacitation* process, as it is named, the female's fluids modify the surface of the sperm, possibly by removing an inhibitor, so that it can interact with the egg.

The sperm which reaches an egg will first touch the *clear envelope* of gel that sur-

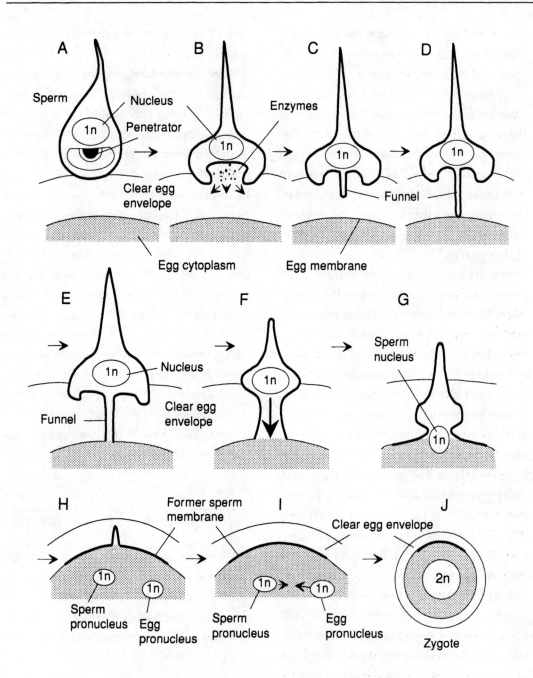

Figure 17 - 1 Stages in fertilization.

rounds the egg. On contact, the *penetrator* (acrosome) of the head of the sperm breaks open and releases enzymes which dissolve away the envelope (Fig.17-1A and B). Next a funnel made of a membrane that comes from the head of the sperm goes through the dissolved gel, reaches the membrane of the egg, and fuses to it (Fig. 17-1C, D, E). The funnel then opens into the egg, and the nucleus of the sperm enters the egg's cytoplasm where it remains to the side away from the nucleus of the egg (Fig.17-1F, G, H).

Once the nucleus of the sperm, now called *pronucleus*, is inside the oocyte, the oocyte starts its second meiotic division to give off another *polar body*. Remaining behind in the fertilized egg is the haploid oocyte having twenty-three chromosomes in its nucleus, and the sperm's pronucleus also having twenty-three chromosomes (Fig.17-1H). The sperm pronucleus loses its own membrane and gets a new one made by the egg's cytoplasm. The sperm pronucleus and the egg's nucleus (now also called a pronucleus) fuse by a process named *syngamy* (Fig.17-1I). The result of syngamy is a diploid nucleus which then forms its own nuclear membrane. At this point the genetic step of fertilization is over, and the *zygote* (fertilized egg) is truly formed (Fig. 17-1J). Next, when the zygote initiates its first mitotic division to commence the development of the embryo, we can say that the second step of fertilization, the *activation* step, has taken place.

Although the foregoing leaves out a great deal, it describes the basic elements of fertilization: (a) the capacitation of the sperm after a certain residency time; (b) the reaction of the penetrator; (c) the fusion of the membranes of the sperm and egg; (d) the entrance of the sperm nucleus into the egg's cytoplasm; (e) the elimination of the polar body of the secondary oocyte during the second meiotic division; (f) the fusion of the male and female pronuclei (syngamy) to form the zygote; and (g) the activation of the metabolic events starting mitosis and development of the embryo. Interestingly, if you inject an entire mature sperm into the cytoplasm of a secondary oocyte, syngamy will usually not occur; the presence of the naked sperm nucleus in the cytoplasm of the egg, however, will lead to the fusion of the two pronuclei.

Variations in Fertilization

A number of variations can occur in the fertilization process. Abnormal variants, described in Chapter 12, including the cases of polyploids and hydatidiform moles, lead to embryos having chromosomal aberrancies.

Development Before Implantation

Once the egg is fertilized in the oviduct, it goes through a number of developmental steps before it embeds itself by a process called implantation into the nutritious inner lining of the uterus (Fig.17-2).

Development of the Blastocyst

Recall that the fertilized egg, the zygote, is a relatively large cell having much more food reserves in its yolk-filled cytoplasm than do most other cells. This food reserve supplies the necessary nutrients and energy to the dividing zygote as it travels along the length of the oviduct until it reaches the uterus. During this journey, the zygote goes through a number of mitotic divisions to form the multicellular hollow ball known as the *blastocyst* (Fig.17-3).

In each of those mitotic divisions, the zygote essentially cleaves itself in half, thereby dividing the cytoplasm equally among the two daughter cells as each new nucleus forms. For example, the two-cell stage formed is about the same volume as the zygote from which it came. Likewise, the

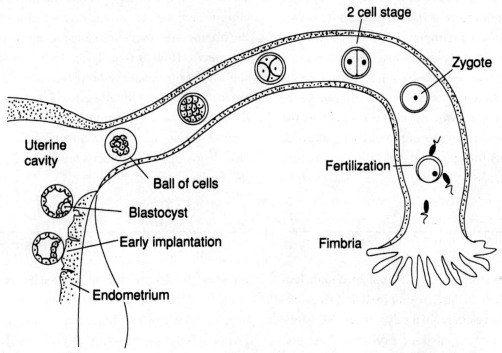

2 cell stage

Zygote

Uterine cavity

Ball of cells

Blastocyst

Early implantation

Endometrium

Fertilization

Fimbria

Figure 17 - 2 Stages of egg in oviduct and uterus from fertilization to early implantation.

four-cell stage, the eight-cell stage, the sixteen-cell stage, etc., are also no larger than the original zygote. This process continues until a solid ball of cells forms, which by the fourth or fifth day after fertilization, starts to hollow itself out to form the blastocyst. Each cell of the one hundred or so cells of the blastocyst is about the same size as the cells found in the adult organism.

Up until this point in development, all these mitotic cell divisions took place while the developing blastocyst was moving down the oviduct towards the uterus. During this whole time the developing blastocyst was surrounded by the clear egg envelope. The envelope serves to hold the cells of the developing blastocyst together and to prevent it from sending out cellular processes and implanting in the oviduct.

Once the blastocyst reaches the cavity of the uterus, it is about out of food reserves. The envelope disappears, and the blastocyst is ready to implant into the lining of the uterus (endometrium) in order to get sufficient nutrients to develop into a mature fetus.

fate, and other cells have a different fate. The blastocyst has two kinds of cells, the single layer on the outside, and a clump of cells at one end of the inside of the hollow ball.

The single layer of flat cells making up the periphery of the hollow blastocyst are called *preplacental cells (trophoblast)*. These cells are the first to touch the inner lining of the uterus and to start the process of implantation. To implant, the outer surface of the preplacental cells sends out numerous small finger-like projections called *microvilli* (Fig. 19-1). The preplacental cells eventually give rise to the cells which form the placenta and certain embryonic membranes.

The other group of cells of the blastocyst, those forming a mass of about thirty cells clumped together at one end on the inside of the blastocyst's cavity, are called as a group the *inner cell mass* (Fig. 17-3). These cells, slightly smaller than the preplacental cells, are the ones that eventually give rise to all of the cells that will make up the embryo and fetus. Most of our study of human embryonic development will focus on the fate of these cells.

Blastocyst

This structure (Fig.17-3), shaped like a hollow ball, is the first differentiated structure that develops from the zygote. By differentiated, we mean that some cells now have one

Twinning

Although we have not studied the development of the embryo yet, now is a good time to take up the subject of twinning. Twins, of course, refer to two individuals

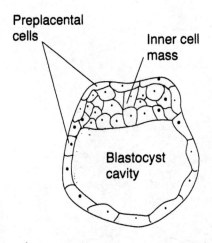

Preplacental cells

Inner cell mass

Blastocyst cavity

Figure 17 - 3 Simple blastocyst.

born at about the same time from the same mother. Twins are usually classified in terms of their genetic makeup. Those having identical genes and derived from the same zygote are called either *identical* twins or *monozygotic* twins, whereas those having different genetic makeups and derived from two zygotes, are called *fraternal* twins, or *dizygotic* twins. The origin of dizygotic twins is quite simple to explain; they are formed when two separate eggs are released by the mother's ovary at about the same time, and each is fertilized by a different sperm. Therefore zygotic twins may be of different sexes, or of the same sex. These twins do not necessarily resemble one another any more than they would other brothers and sisters born of the same parents years apart. When we later study the formation of

the placenta (Chapt. 22), it will become apparent that dizygotic twins each develop in the mother's uterus with separate placentas.

Although monozygotic twins form less frequently than do dizygotic ones and are derived from a single fertilized egg, they can originate by a number of different pathways. In some cases, the cells of the two-cell stage zygote separate from one another and each develops into a separate embryo in much the same way as do dizygotic twins, each with their own placenta. The monozygotic twins developing in that manner, however, are genetically identical in contrast to dizygotic twins which are not.

Monozygotic twins can arise in two other ways, both of which involve the splitting in two of the inner cell mass of the blastocyst (Fig. 17-4). In the more common of the two cases, the blastocyst will consist of a single outer trophoblast layer and two separate inner cell masses. Each of the cell masses will develop within the same outer placenta, but within their individual protective amniotic sac (Chapt. 18). In the rarer case of the two, the inner cell mass spreads itself out over a wider area and does not separate completely until a little later. As a consequence, two individual embryos develop, but in this case within the same amniotic sac and same placenta. Is it possible to have identical triplicates? How?

Is it possible to have "identical" twins of different sexes? The obvious answer is "no,"

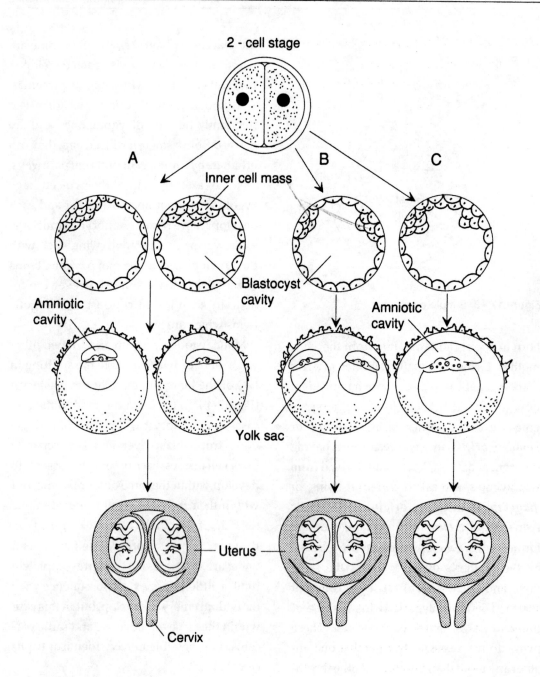

Figure 17 - 4 Different ways to form monozygotic (identical) twins.

but nature does not always work in obvious ways. Try to recall the material that you read in the chapter on chromosomal aberrations. With that information, you should be able to work out a situation in which it is possible to have monozygotic twins of the opposite sex, and you should be able to specify certain characteristics of one of the twins.

While on the subject of the development of a single fertilized egg, I would like to mention an unusual variation that will not give rise to monozygotic twins, but to an embryo in which half of its cells contain genes which came from one sperm, and the other half contain genes coming from a different sperm. In this case, which usually originates in an egg that was released by the ovary of an older woman, the primary oocyte does not give off a smaller polar body when forming the secondary oocyte. Instead, in Meiosis I the primary oocyte gives rise to two secondary oocytes of about equal size which remain stuck to one another. When this "two-celled" oocyte is released at ovulation, each of the oocytes can be fertilized by a different sperm. The resultant doubly fertilized "two-cell zygote" will then develop into a single embryo within the same amnion and placenta. This embryo, however, will develop into a mosaic with half of its cells possessing genes from the mother and one sperm and the remaining cells containing genes from the mother and from a different sperm.

Chapter 18

Implantation, the I.U.D., and Test Tube Babies

Implantation refers to the process by which the blastocyst becomes embedded into the inner wall of the uterus where it will gain nutrition and develop. In this chapter, we will take up four aspects of implantation: (a) the steps in normal implantation; (b) ectopic implantations (i.e. implantations taking place at some site other than the uterus); (c) prevention of implantation by means of the I.U.D.; and (d) the artificial implantation of blastocysts as is done in the cases of "test tube babies."

For normal implantation to take place, the conditions in the body must be just right. Implantation is timed to take place about five to seven days after ovulation occurs, and hence at the same time that the corpus luteum is releasing progesterone and estrogen. During that period the zygote develops into the blastocyst and is carried along the oviduct by means of contractions of the oviduct and the movement of the little hair-like cilia lining it. Immediately before the blastocyst reaches the uterus, the clear envelope surrounding it ruptures and is lost. The preplacental cells on the outer periphery of the blastocyst are ready to implant if the lining of the uterus is prepared, by the secretion of progesterone and estrogen, to accept the blastocyst . It is this latter step, the acceptance of the blastocyst by a prepared uterus, that is crucial if implantation is to take place.

Normal Implantation

First the blastocyst free of its clear egg envelope reaches the upper uterine cavity (Fig.18-1A). Once those preplacental cells adhering to the inner cell mass touch the inner lining of the prepared uterus, their microvilli begin to invade the wall of the uterus. Starting at the time that the blastocyst is embedding itself into the uterus, the preplacenta cells start to differentiate into two layers: the *invasive preplacenta* (also called syncytiotrophoblast) and the *noninvasive preplacenta* also called cytotrophoblast) (Fig.18-1B, C, D). The invasive preplacenta, which starts to form from the preplacental cells adhering to the inner cell mass, eventually grows to surround the noninvasive preplacental cells. While the cells of the invasive preplacenta are differentiating,

they lose the membranes surrounding their individual cells and the nucleuses left behind become surrounded by one large cell membrane. It is this invasive preplacenta that: (a) invades the lining of the uterus; (b) engulfs the maternal blood vessels in the uterus; (c) eventually grows around the noninvasive preplacenta and the whole embryo; (d) secretes progesterone and estrogen; and (e) finally ends up as the major part of the placenta of the fetus.

While all these events are happening, a distinct *noninvasive* preplacenta of a thickness of one cell layer is retained beneath the invasive one and around the entire developing embryo (Fig.18-1D). It is this noninvasive preplacenta which starts to secrete the *human chorionic gonadotrophin* (hCG) that functions to keep the corpus luteum actively secreting progesterone and estrogen during pregnancy. If no hCG were secreted, then the *corpus luteum* would not be signalled that implantation had taken place, it would degenerate to form the corpus albicans, and menstrual bleeding would take place, thereby eliminating the implanted embryo along with the lost inner lining of the uterus. Recall that it is this hCG released by the noninvasive preplacenta which serves as the basis for home pregnancy test kits.

Other changes occurring simultaneously in the blastocyst during implantation, such as formation of the amniotic cavity and the yolk sac, are described in Chapter 19.

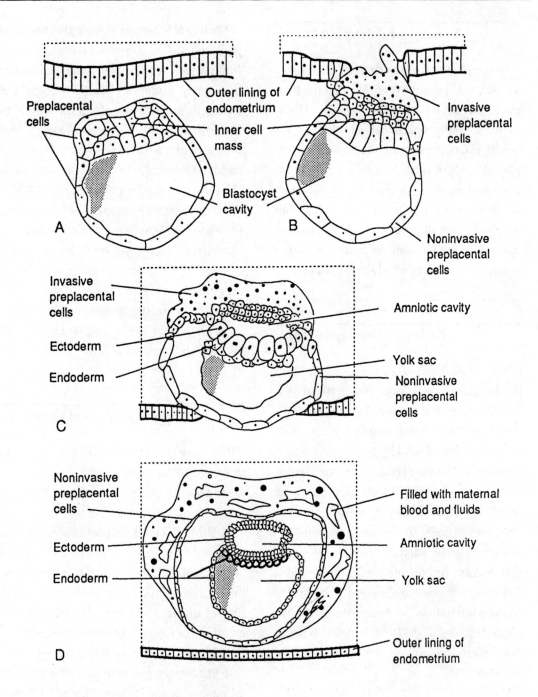

Figure 18 - 1 Implantation of blastocyst and formation of amniotic cavity and yolk sac.

Successful Implantation

For implantation to be successful, the blastocyst has to be entirely enclosed in the uterine tissue, but it must not embed too deeply there. Secondly, spaces filled with maternal blood must appear in the invasive preplacenta; these provide ample nutrition to the growing embryo. And lastly, the menstrual bleeding has to be prevented from occurring. All these events are dependent upon the correct balance of hormones in the prospective mother's body at the time.

Ectopic Implantations

Approximately one or two percent of all implantations are *ectopic*, that is the blastocyst implants somewhere else other than in the uterus. Most ectopic implantations occur in the oviduct. Possibly the egg envelope ruptures before the blastocyst leaves the oviduct, and the trophoblast cells start to embed there. When these implantations develop into ectopic pregnancies, serious complications may arise, such as massive bleeding, that require immediate medical attention. Other symptoms are severe abdominal pains and failure to menstruate. In the United States and Great Britain, ectopic pregnancies account for ten percent of maternal deaths.

There are cases in which the blastocyst goes the wrong way and falls out of the fimbrial end of the oviduct into the abdominal cavity. Some of these blastocysts may embed in the ovary or along the wall of the abdominal cavity. In recent years, a child was born of a woman whose blastocyst embedded ectopically in the abdominal cavity close to a rich supply of blood. The child had to be extracted by making an incision into the abdominal cavity, not the uterus as in a Cesarian delivery. Although it was healthy in every respect, such a successful ectopic pregnancy is an extremely rare event — but possible. Interestingly, it is experimentally possible to successfully implant mouse embryos in the scrotum of mice.

Delayed Implantations

There are two kinds of delayed implantations: those that can occur naturally and those that are carried out artificially for either commercial or medical reasons.

Natural delayed implantations do not occur in humans, but rather in such animals as marsupials, badgers, seals, and sea lions, in which many young are born at the same time of the year — a time favorable for the newborn, or a time when the mates are there to help care for the growing young. In the case of badgers, for example, after the blastocysts reach the uterus, they may "hibernate" there from periods of a few months up to two years before implanting. Obviously, the lining of the uterus is not prepared to accept those blastocysts until some event, probably seasonal, triggers the organism's hormonal system.

Artificially Delayed Implantations

There are at least two obvious reasons to devise ways to delay implantations artificially. One is a commercial reason devised by the animal breeding industry. The other is related to the new human technology involved with in vitro fertilization.

With regard to the animal breeding example, what if a rancher had a prize cow in Montana, and another rancher in Australia wanted to raise progeny of that prize cow on his own farm. It could be done by washing the blastocysts out of the uterus of the prize cow, and then placing the blastocysts into the oviduct of a rabbit where it will not implant, but can survive for some time. The bunny can be shipped rather cheaply to Australia where the blastocysts can be removed and implanted into a cow whose uterus has been made receptive through hormone injections. Now it is even been possible to ship overseas frozen animal blastocysts which can be thawed at their destination and successfully implanted there.

"Test Tube Babies," or *In Vitro* Fertilization

In vitro fertilization is a technique whereby it is possible to withdraw an unfertilized egg from a woman's follicle, fertilize it outside of the body with sperm, and replace the resultant blastocyst in the same woman's uterus. If the blastocyst implants successfully, then she will most likely deliver a normal child in eight and one half months. In essence, through the process just summarized above, the woman bypasses those steps that usually take place in the oviduct — fertilization and development of the blastocyst from the zygote. Therein lies the main reason for developing the technology of *in vitro* fertilization — to allow those women with blocked oviducts to bear their own children.

As indicated earlier in this chapter, the key puzzle to be solved before successful *in vitro* fertilization could be achieved was that of implantation. The technology of the other major steps of *in vitro* fertilization had been perfected years before. Successful implantations were a problem because of side effects resulting from inducing ovulation. To make certain that the physician could find a follicle in the ovary which was ready to ovulate, the woman had to be given hormones to stimulate the follicles to mature and to ovulate. By applying those hormones, however, the delicate balance of hormones, especially the ste-

241

roid ones needed to prepare the uterus for implantation, was usually upset. Thus, the last major technological feat to be mastered was to synchronize the readiness of the blastocyst and of the uterus. There now are a number of ways to accomplish this coordination, one of which, for example, is to freeze the blastocyst and to implant it during a subsequent menstrual cycle when the woman's uterus is naturally prepared to accept it.

The individual steps of *in vitro* fertilization should now be understandable to you if you have assimilated the material covered thus far in this book. They are: (a) The woman is given a hormonal treatment to stimulate ovulation(s). (b) Almost thirty-four hours later the egg is removed by the aid of a *laparoscope,* a sort of "periscope" which allows the physician to view the inside of the abdominal cavity where the ovaries are situated. The egg (actually the secondary oocyte) is collected by suction applied through a syringe attached to the *laparoscope,* the needle of which is inserted into a ripe follicle. (c) The patient is then treated with a regimen of estrogen and progesterone to prepare her uterus for implantation. (d) Sperm, collected before the egg is taken from the follicle, is washed, diluted, and put in a solution that simulates the fluids of the oviduct, where fertilization normally takes place. In that solution, the sperm undergo the process of capacitation, which is necessary for fertiliza-

tion to occur. (e) The sperm and egg are placed together in a solution kept in a dish free of bacteria and other contaminants; there fertilization occurs outside of the body, i.e. in glass, which is what *in vitro* means in Latin. (f) The fertilized egg is carefully monitored for four to six days as it develops into a blastocyst. (g) The blastocyst is either frozen for future implantation, or it is immediately put into a fine flexible tube which is inserted into the upper uterus where the blastocyst is released. (h) If the uterus is prepared, the blastocyst will implant and the resultant embryo will develop into a healthy child.

In the early nineteen eighties, births by *in vitro* fertilization started to become relatively common. Now that the number of such "test tube" babies is bound to increase, serious attention is being paid to legal and moral problems that may arise, as well as to possible abuses of the procedure. For example, what are the rights of the child should the procedure lead to birth defects? What are the implications of implanting the blastocyst in a *surrogate mother*? Will such be the way for the rich, or for tyrannical governments which could use surrogate mothers simply as slaves to bear the genetic offspring of the leadership?

On the more positive side, *in vitro* fertilizations offer a number of opportunities for infertile couples to conceive children, to prevent birth of defective children, and even to eliminate certain genetic diseases. As an example, use of *in vitro* fertilization and of a surrogate mother can allow couples to have their own genetic offspring who otherwise could not because of the female's damaged uterus or because the uterus had been removed in a hysterectomy. Surrogate mothers could also be used by a mother who finds that her herpes infection might damage the child.

Other couples might choose to use *in vitro* fertilization so that some cells of the blastocyst can be removed and tested. For example, parents harboring sex-linked genetic diseases realize that if the child is a boy, then the chances are fifty-fifty that he would carry the genetic defect. Such parents may use the Barr test on a blastocyst cell and decide not to implant it if the Barr test shows it to be male. Likewise, older women may chose to have cells of the blastocyst analyzed for their karyotype. Should the cell show three chromosomes of the number twenty-one type, thereby indicating the child will have Down's syndrome (Trisomy 21), the parents will have the option of deciding whether or not to implant the blastocyst. Thus, by using tests before implantation of the blastocyst that are normally reserved for amniocentesis done in the early months of pregnancy, the parents might avoid some of the difficult decisions and psychological and physical trauma associated with later abortion of a defective embryo.

Even more dramatic possibilities will exist in the near future with the development of genetic engineering. Just as it is possible to insert mammalian genes into bacterial cells, genes that will become a permanent part of the genetic makeup of those bacteria, it should be possible to insert human genes into the unfertilized egg or the zygote after the egg has been removed from the mother. Just think what that would mean to parents carrying such genetic defects as sickle cell anemia, galactosemia, phenylketonuria, or Tay-Sachs syndrome. By inserting the correct gene that was missing in the parent(s), the child could make the missing protein or enzyme and neither it nor its future offspring would have to suffer from the genetic defect that has plagued its ancestors.

Prevention of Implantation — the I.U.D.

The intrauterine device, i.e. the I.U.D., is a birth control device that does not act by interfering with ovulation, as do oral steroid contraceptives; there is no evidence, for example, that I.U.D.s cause any significant changes in the female's hormonal function or in the menstrual cycle. Nor does the I.U.D. act by preventing sperm from reaching the egg, as do the diaphragm and condom. Instead, the I.U.D. is a foreign object which is placed in the uterine cavity for long periods of time, and as a result prevents the blastocyst from implanting.

Mode of Action

The exact mechanism by which I.U.D.s interfere with implantation in human beings is not known. Although they may interfere with implantations in a number of ways, however, we are not sure which is the major mechanism.

For one, I.U.D.s alter the cells that line the uterus and their biochemical reactions. They act as a foreign body, thereby stimulating an inflammatory reaction. After the I.U.D. is inserted in the uterus, many white blood cells are found there. White blood cells normally swallow foreign materials, such as bacteria and viruses. Current theory has it that the large concentration of these white blood cells actually consumes the blastocyst and/or the sperm.

Most of the I.U.D.s that act in the above manner are made of inert plastics. Others are constructed to release toxic copper ions that may poison the blastocyst. Still others may release steroid hormones that affect the endometrium deleteriously.

Effectiveness

Research shows that I.U.D.s are "a generally safe, effective and useful form of birth control." In addition, I.U.D.s greatly lower the rate of maternal mortality, especially in some developing countries where up to three hundred mothers per one hundred thousand die by giving birth; only one to three per one hundred thousand die of those using I.U.D.s.

Compared to other methods of contraception, the I.U.D.s rate extremely well. One study showed that the failure rate of women using I.U.D.s to prevent pregnancy to be 2.9 per 100, compared to 2.0 for oral contraceptives, 6.6 for condoms, and 10.3 for diaphragms. Worldwide, it is estimated that sixty million I.U.D.s are in use.

Advantages and Disadvantages

I.U.D.s have advantages over other contraceptives, other than their effectiveness. They require a single insertion and give long-term protection. Nevertheless there still are some problems with their use to be recognized and solved. For one, they lead to increased menstrual bleeding, sometimes twice the volume that is normal. Secondly, if not inserted properly and by experienced personnel, they can be expelled soon after being inserted. Finally, they lead to an increased frequency of pelvic infections. Among young sexually active women, one form of pelvic inflammatory disease can be sexually transmitted. Fourth, severe cramps are sometimes experienced by I.U.D. users. These may be caused by the uterus contracting to expel a foreign body from its midst. Finally, relatively more ectopic pregnancies occur among I.U.D. users than among users of other forms of contraception. Estimates indicate that 1 out of every 30 pregnancies of I.U.D. users is ectopic compared to 1 out of 250 in the general population. The chance of ectopic pregnancy appears to be reduced, however, if the I.U.D. is removed as soon as the pregnancy is detected.

Prospects and Choices

Because I.U.D.s are so effective in preventing conception, research on improving them still goes on. Some of the research aims to develop I.U.D.s that: (a) can be used for about ten years; (b) have compounds attached to them which reduce bleeding; (c) are designed to reduce the chances of their being expelled; (d) have an improved fit; and (e) can be inserted easily, safely, and relatively permanently by the user without much training.

Over the past five years there has been some concern regarding the use of I.U.D.s because of some multimillion dollar legal suits that have made the headlines. It now seems, however, that I.U.D.s are once again being discussed as a major effective birth control device.

It is not my intent to recommend any particular contraceptive method discussed thus far, such as oral contraceptives, I.U.D.s, condoms, or vasectomies and tubal ligations. I try to explain only how they work, and their advantages and disadvantages.

Chapter 19

Early Development of Human Embryo

By definition, the development of the egg from fertilization to birth goes through three periods. The first, from fertilization to the fifteenth day, entails the formation of the *blastocyst*. The second, the *embryonic* period, from the third week through the twelfth week, takes the blastocyst through the developmental stages that give the embryo an unmistakably human appearance. In the last, the *fetal* period, fourth month to birth, the developing child-to-be increases in size many fold, its organs develop more finely so that they will eventually function independently of the mother, and the shape and proportion of the parts of the body assume those required for a newborn infant to survive.

The events taking place in the embryonic period are much more complex than are those of the blastocyst and fetal periods. It is in the embryonic period that the fate of the various cells and layers of cells is determined, that the tissues and organs first form, and that the shape of the human body is established. Because of the complexity of these steps and the necessity for integrating the timing of the onset of each step, there are many points in the development of the embryo for things to go wrong. For it is in the embryonic stage that such externally supplied factors as drugs, alcohol, food additives, diet, radiation, smoking, hormones, and viruses can interfere with normal progress. Thus I place special emphasis in this chapter on explaining the initial crucial processes in the development of the embryo. If prospective parents were aware of the complexities involved in the development of the embryo, I believe they would take extraordinary pains to insure that during the first three months of pregnancy the embryo not be exposed to harmful external factors that might cause birth defects (Chapt. 23).

In the present chapter, we will cover four aspects of early embryonic development: (a) the fate and functions of the preplacental cells; (b) the development of germ layers; (c) the immediate fate of these germ layers; and (d) the formation of the amnion and the yolk sac.

Fate and Roles of the Preplacental Cells of the Blastocyst

We have already discussed the role of the preplacental cells during implantation, and their differentiation into the invasive preplacental cells, situated on the outside of the blastocyst and the noninvasive preplacental cells, situated between the invasive ones and the developing embryo. A major function of the invasive preplacental cells is to send out many fine villi into the maternal tissue in order to increase the surface area of contact between embryonic and maternal tissues, thereby increasing the exchange of nutrients and wastes (see Chapt. 22).

As the embryo develops, both layers of preplacental cells take on major roles in producing hormones. The invasive preplacental cells start very early in their formation to secrete the steroids progesterone and estrogen. Because the developing placenta of the early embryo has relatively few cells compared to the larger placenta of the fetus, the steroid hormones that it produces have much less effect on the physiology of mother than does the larger amount of those hormones secreted by the corpus luteum of the ovary. On the other hand, the steroid hormones released by the preplacental cells, and later by the placenta, are in direct contact with the inner lining of the uterus and take on a greater role in helping to maintain it as the pregnancy progresses.

Eventually, toward the end of the third month of pregnancy, the placental cells are producing significantly more progesterone and estrogen than the corpus luteum does, and the corpus luteum, consequently, takes on a less significant role in maintaining the uterus at about this time. Now you can understand why removal of the ovaries during early pregnancy will cause the embryo to abort. It will not abort if the ovaries are removed after the third month because the placenta itself will provide sufficient steroid hormones to maintain the inner lining of the uterus.

Recall that progesterone also reduces the excitability and contractility of the uterine muscles, and in so doing, helps to maintain pregnancy by preventing the potential expulsion of the embryo as a result of uterine contractions. The estrogen produced by the placenta, in addition to maintaining the inner lining of the uterus, also promotes the development of the breasts and uterine muscle.

The primary hormone produced by the noninvasive preplacental cells is *human chorionic gonadotrophin* (hCG). A major function of hCG is to stimulate the maintenance of the corpus luteum during pregnancy. This hormone, which serves as the basis of the pregnancy test, is also thought to stimulate the buildup of protein in the embryo.

Development of Inner Cell Mass — An Overview

All of the tissues and organs of the embryo as well as the cells of the temporary embryonic cavities, such as the amnion and yolk sac, develop from the inner cell mass of the blastocyst. Because the events leading to the formation of those tissues, organs, and cavities are so crucial to our understanding of embryonic development, I will first present a few basic definitions. Then I will describe the individual events one by one, and finally I will integrate all those facts to give a comprehensive picture of how these events are interrelated.

Definitions

The *amnion* (amniotic sac) and the *yolk sac* (Figs. 18-1 and 19-1) are the two temporary embryonic cavities that we will consider. The amnion and its fluids protect the embryo and serve as a medium from which the embryo and fetus get some nutrition and into which the embryo and fetus expel their liquid and solid wastes. This is the cavity from which fluids can be withdrawn during amniocentesis. The yolk sac, part of which becomes a portion of the gut, serves mainly to supply food reserves to the early embryo. For a short time it is a site for manufacturing

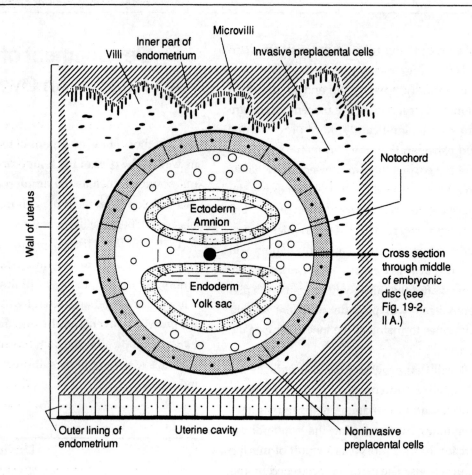

Figure 19 - 1 Development of amnion and yolk sac (seen in cross section, i.e. thin slice through middle).

red blood cells for the embryo. The yolk sac plays a relatively less significant role in human embryos compared to its role in the embryo of birds and reptiles. With those animals, development takes place within the confines of the shell of an egg, and thus the yolk serves as the main source of food and consequently is quite large.

The interface between the early amnion and the yolk sac consists of a layer of *ecto-dermal* cells (ectoderm) on the amnion side and a layer of *endodermal* cells (endoderm) on the side of the yolk sac. Furthermore, ectoderm surrounds the entire amniotic cavity and endoderm surrounds the entire yolk sac (Fig. 19-1). A third middle layer of cells, called *mesoderm*, develops from the ecto-derm and endoderm at the junction between those layers, and always lies between them (Fig. 19-2, row III).

These three cell layers of the early embryo are called the *primary germ layers* and make up the *embryonic disc* (Fig. 19-2, row 1) from which the embryo develops. The ectoderm gives rise to all the cells of the brain, spinal cord, and nervous system, and also to the cells of the outer layer of skin. The endoderm develops primarily into the lining of the gut (esophagus, stomach, intestines, and rectum), whereas the mesoderm gives rise to virtually all other tissues and organs such as muscle, bones, and gonads and those of the circulatory system.

A fourth temporary embryonic structure, which is not composed of cells but which is formed in the mesodermal area, is the *notochord*. This rod-like structure (Fig. 19-2, row II) is relatively stiff and later becomes incorporated as part of the vertebrae (backbones). The notochord is located halfway between the two equal sides of the elongated germinal disc running lengthways, and it gives the embryo a sort of *bilateral symmetry*, that is, each side of the embryo is almost a mirror image of the other.

Formation of the Amnion and Yolk Sac

While the blastocyst is implanting into the uterine wall and the preplacental cells are differentiating into the invasive and non-invasive layers (Figs. 18-1 and 19-1), the inner cell mass becomes hollowed out and forms the ectoderm-lined amnion; meanwhile the blastocyst's cavity is gradually pushed aside by the endoderm cells which surround the expanding yolk sac. Immediately between the amnion and the yolk sac is the three-layered embryonic disc consisting of the ectoderm on the amnion side, the endoderm on the yolk sac side, and in between, the mesoderm with its notochord.

Development of the Neural Tube

To appreciate fully these dynamic steps of development, you have to think three dimensionally when viewing the figures in this chapter even though most of them are two-dimensional representations. In addition, to visualize the neural tube forming from the ectodermal layer of cells, I suggest you imagine that you are a tiny creature swimming around inside the amniotic cavity, and are looking down upon the ectodermal top side of the embryonic disc (Fig. 19-2, row I).

Thus, while viewing the top ectodermal surface of the embryonic disc, you will eventually see the first indication of a neural tube forming. First you will see a shallow, thin groove that starts to elongate and extend towards both ends of the disc. When the groove reaches about four fifths of the way toward the forward end, it begins to widen, while the sides of the groove at the initial end of the furrow starts to fuse (Fig.19-2, row I).

TIME

A   B   C   D

Row I: View of embryonic disc from " the top. "

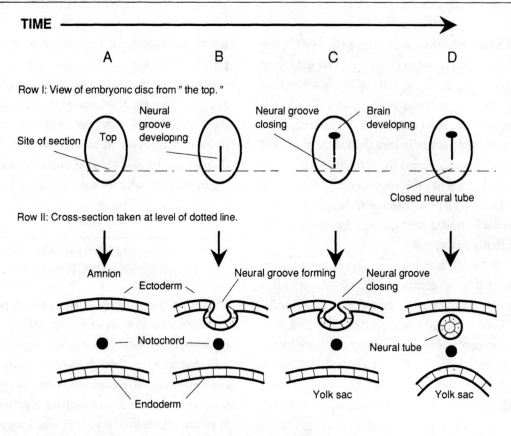

Row II: Cross-section taken at level of dotted line.

Row III: While the neural tube is forming, the mesoderm is also forming. Clumps of mesoderm form the somites, A new cavity, called coelom, or body cavity, forms in between the mesoderm cells away from the somites.

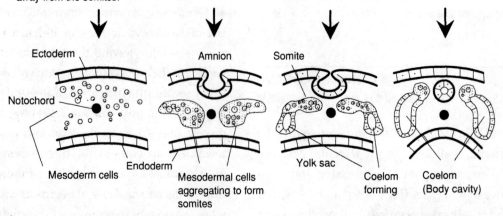

Figure 19 - 2 Development of embryonic disc, germ layers, notochord, neural tube, somites, and coelom.

What actually is happening is that the ectoderm of the embryonic disc is *invaginating* to form a neural groove which soon thereafter begins to seal itself to become the neural tube and, eventually, the spinal cord. The part that widens at the forward end eventually becomes the brain.

To get a clearer view of the invagination process, look at Figure 19-2, row II, which shows only the ectoderm, endoderm, and notochord of the disc. Think of the embryonic disc and its notochord as a loaf of rye bread (embryonic disc) having a steel rod (notochord) inside along its length, and each cross section that you are looking at is a slice of that bread with the crust on top as ectoderm, and the crust on the bottom as endoderm. Put all the slices (i.e. cross sections) back together and you get the complete loaf (i.e. the germinal disc). Note in Row II of Figure 19-2 that the invagination starts in B (along the top crust just over the steel rod), begins to seal over to form a neural tube enclosed in ectoderm cells in C, and completely seals over to form a tube which has become separated from the outer surface of the ectoderm in D.

Formation of the Mesoderm and Somites, and Beginning of Body Cavity

To get an idea of how these structures form, focus on row III of Figure 19-2. Early in the development of the germinal disc (see A), the mesoderm cells slough off the multiplying ectodermal and endodermal cells. Some of the mesodermal cells start to form the notochord very early. As the groove of the prospective neural tube forms along the ectoderm, some of the mesodermal cells compact themselves together to form structures called somites. The somites arrange themselves in pairs along the length of the embryo's notochord (see Fig. 19-2, row III, B); they are also represented in Figure 19-4, where they appear as "bumps" under the skin. Eventually the main body of the somites will give rise to the vertebrae around the spinal cord and to the muscles situated there.

The tail ends of the somites split apart and each forms a cavity (Fig. 19-2, row III, C) which gets much larger while the neural tube is forming (Fig.19-2, row III, D). This cavity eventually gets even larger (Fig. 19-3) as the embryo develops. Eventually the cavities formed from each somite fuse to develop into a single body cavity (abdominal cavity, coelom) in which are found all the internal organs formed from the mesoderm, such as the heart and kidneys.

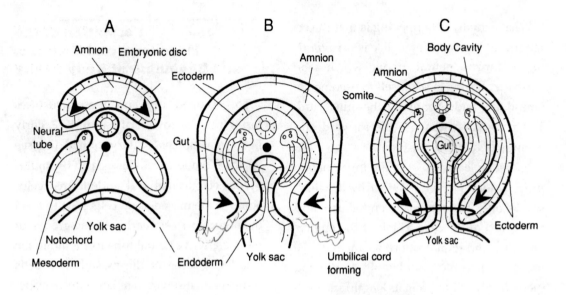

Figure 19 - 3 Further development of amnion and yolk sac and formation of gut from endodermal cells of yolk sac (arrows indicate direction of tissue growth).

Enlargement of the Amnion and Body Cavity, and Development of the Gut from the Yolk Sac

Although we have focused thus far only on the development taking place in the embryonic disc, you must remember that the disc is bounded by the amnion on its ectodermal side, and by the yolk sac on the endodermal side. To give you a better picture of the dynamics of how the amniotic cavity and part of the yolk sac develop, Fig. 19-2, row III, D is redrawn in Figure 19-3A to include the ectodermal layer of cells that also surrounds the amnion, and the endodermal cells that also surrounds the yolk sac.

The two dark arrows at each side of the amniotic cavity in Figure 19-3A indicate that as the embryo develops, the amnion gets larger and moves in the direction of the arrows, thereby "pushing" the embryo which is developing within the disc downwards and finally inwards (Fig. 19-3B). While the amnion is expanding as indicated by the arrows in Figures A and B, the endoderm of the yolk sac is starting to form an additional cavity; that cavity eventually becomes sealed off, as Figure 19-3C indicates, and becomes the inner lining of the gut of the embryo. Depending upon where the cross-section was made, the gut illustrated could represent the esophagus, stomach, intestine, or rectum. At the same time the gut is being

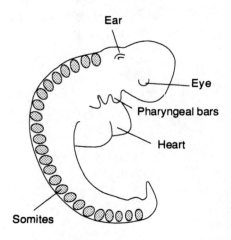

Figure 19 - 4 View of somites in 27 day embryo. (Somites appear as "bumps" showing through skin; notochord not shown.)

formed, the remaining endodermal layer of cells and its enclosed yolk sac is pushed outside of the developing embryo (Fig. 19-3B and C), and it becomes displaced against the placental tissues. The yolk sac there still supplies some nutrition to the early embryo, but its primary function is to manufacture red blood cells for the embryo.

While the amnion is enlarging, the gut forming, and the yolk sac being displaced, the body cavity surrounded by the mesodermal extensions of the somites begins to enlarge (Fig. 19-3C). By the time the complete gut forms, the mesodermal lined body cavity begins to occupy most of the space of the embryo, except for that occupied by the gut, neural tube, and somites. Not shown in these figures are the organs of mesodermal origin that become suspended in the body cavity.

Note that the outside layer of the gut eventually is derived from mesodermal cells, which will become the muscular layer of those organs.

The narrow connection between the embryo and the placenta (Fig. 19-3C), a connection which gradually develops into the shape of a tube and becomes surrounded by the amniotic cavity, is the *umbilical cord*. It is through this umbilical cord that the blood vessels traverse which connect the circulatory system of the embryo and fetus to the wall of the placenta where an exchange of nutriments, gases, and wastes take place.

A Vestigial Cavity — the Allantois

Although the allantois (allantoic cavity) plays no major role in the human embryo, it does develop slightly. I mention it only as it represents an interesting vestige of vertebrate evolution. This cavity, which grows out alongside the yolk sac, plays an important role, however, in embryos of birds and reptiles. Recall from your general biology that both birds and reptiles develop in eggs that are deposited outside the body. The developing embryos of birds and reptiles, therefore, would have a problem of getting rid of their toxic waste products that, in human embryos, are gotten rid of through the placenta into the mother's circulatory system and kidneys. Birds and reptiles have solved this problem by using the allantoic cavity as a "dumping site" for their toxic wastes. Instead of making soluble urea to pass into the mother's blood stream as mammalian embryos and fetuses do, birds and reptiles make crystals of insoluble uric acid which are pushed aside into the allantois where they are stored in a non-toxic form and do not harm the development of the birds and reptiles.

That human embryos even possess a vestigial allantois serves to remind us that man has descended during evolution from ancestors related to birds or reptiles. As we study the further development of the embryo and of the fetus in the next chapter, we shall see other evolutionary similarities between humans and our predecessors.

Chapter 20

Development of the Embryo and the Fetus: an Overview

It may seem unusual that I have devoted a whole chapter to the blastocyst stage, which lasts only about seven to ten days, and another full chapter to the development of the germ layers and early embryonic cavities, which lasts an additional ten to twelve days. On the other hand, this one short chapter treats the entirety of the embryonic and fetal stages, which occupy a little more than eight months of human gestation. There are some good scientific as well as pedagogical reasons for my seemingly unequal treatment of these various stages in the development of the child.

First, because of the relatively simplicity of the blastocyst, germ layers, and embryonic cavities, it is much easier to discern the differentiation of cells and overall development of the fertilized egg right up to the formation of the early embryo. Secondly, once these initial changes have taken place, the foundation is set for most of the developmental events that follow.

In the *embryonic stage*, i.e. from about two weeks to three months after fertilization, the embryo forms the body's main internal organs, such as the heart, lungs, and stomach, and external appendages, such as the limbs. Even though by the end of the three months the embryo is only about one and one-half inches long, it has already achieved an unmistakably human appearance. Thus, most of the complex development of the child takes place in the embryonic stage, whereas in the *fetal stage* we see: (a) a great deal of growth; (b) the finer development of the internal organs; and, (c) the alteration in the shape and proportion of the parts of the body.

Because of the complexities involved in the formation of the embryo's organs and major body structures in the embryo, there is a greater chance that harmful environmental factors can interfere with their development during the first three months of gestation. Such interference can lead to birth defects in the developing child. By harmful environmental factors, we mean those which enter the body from the outside, such as drugs, food additives, alcohol, cigarette by-products, high doses of vitamins, certain hormones, chemicals found in the work place, some viruses, and various forms of radiation.

Because a major concern of this book is to increase understanding of birth defects and their avoidance whenever possible, and because it is during the embryonic period that most birth defects are induced, I will focus primarily on the general development of the embryo rather than on fetal development. Furthermore, because the fetal period is developmentally less complex, being concerned primarily with growth and reshaping of the body's structures, and less subject to those harmful factors that cause birth defects, I will but barely touch upon this six month period in the development of the child.

At the end of this chapter, I will cite a few general examples of changes in the physiology and behavior of the fetus as it matures and grows, the means whereby the fetus' circulatory system is prepared for breathing air, and some features of the full term fetus. The detailed embryonic and fetal development of one system, the reproductive system, will be discussed in Chapter 21.

Stages in the Development of the Blastocyst and Embryo

The blastocyst and embryo are categorized by embryologists as going through twenty-three stages called "horizons." This is a term borrowed from geologists and means an epoch, or a period of time covering a number of events. Each horizon is not a fixed or static stage of development, but rather is identified with certain distinguishing and recognizable features of the developing blastocyst, embryo, and fetus. In Chapter 13 we have discussed the first seven and part of the eighth, ninth, tenth and eleventh horizons. Those horizons, occurring mostly in the blastocyst and early embryonic stages, are:

Horizon 1.

>One-cell fertilized egg.

Horizon 2.

>Many-celled stage (without a cavity).

Horizon 3.

>Free blastocyst (with a cavity).

Horizon 4.

>Implanting blastocyst.

Early Embryonic Stages

Once the blastocyst starts to implant, a number of dramatic changes occur before a distinct embryo takes shape. These events, described in the previous chapter, occur in:

Horizons 5.-7.

>Development of villi, yolk sac, amnion.

Horizons 8.-9.

>Notochord, neural groove and first somites appear. Heart begins to form.

Horizons 10.-11.

>Gut and up to twenty somites form; heart beats.

Embryonic Stages

The changes that occur in order to form the fully developed embryo are quite dramatic and complex. It will be impossible, therefore, to cover the details involved in the formation of all the organs and structures that develop so perfectly in this two and one-half month period. I suggest that as you study the highlights of embryonic development, you try to imagine all the changes that the original cells of the inner cell mass must undergo in order to form the completed embryo. The cells of the inner cell mass multiply and differentiate

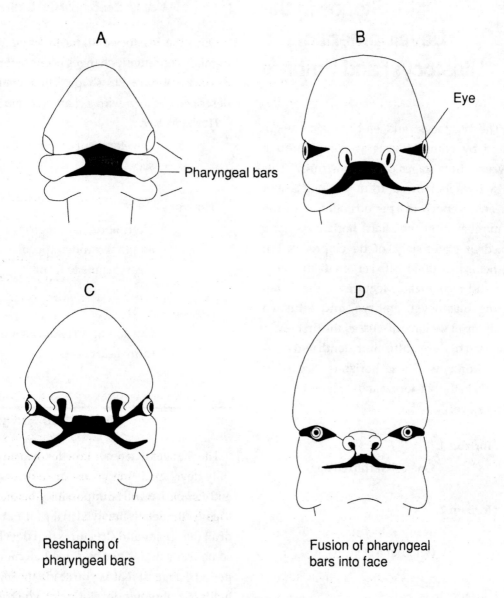

Figure 20 - 1 Development of the human face. A, face of a 4 week embryo. B, face at 4 1/2 weeks. C, at 5 weeks and D, at 7 weeks.

into muscle, nerves, skin, blood, and other types of cells. As the cells multiply and change, they move in groups, known as *tissues*, until they mold the shape of the organs which they eventually form. All of the tissues and organs which form are integrated to work together by the developing nervous, hormonal, and circulatory systems. Remember, all of the information responsible for these changes is found in the forty-six chromosomes of the fertilized egg — one single cell. Here is what happens:

Horizon 12

In this stage, the embryo is twenty-six days of age and about 0.2 of an inch long (Fig. 20-3). The embryo has three bar-like structures at its head end, that is, where the brain is located. These structures *(pharyngeal bars)* eventually form the face of the embryo. The older the embryo gets, the more complex the bars become. The complete sequence of changes taking place as the bars eventually mold into a face is shown in Fig.20-1. From examining these drawings, it becomes apparent why some substances that interfere with the embryonic development of the bars may lead to the formation of a cleft palate. Interestingly, in the embryo of fishes, the bars develop into their gills.

Horizons 13-14

By the time the embryo reaches these stages, at about twenty to thirty days, it ranges from 0.2 to 0.3 inches in length (Figs.

20-2,3). Some blood vessels and a small heart are relatively well developed. Small beginnings of the arms and legs, called limb "buds," can be seen. The ears and eyes start becoming defined, and the lungs start to form.

Horizons 15-16

Now, at thirty to thirty-three days, with the embryo reaching just under a half inch in length, more development of the head can be noticed. It is significantly larger as the brain develops further. Nostrils can be seen forming on the face. As the leg buds grow, the thigh, leg and foot regions can be seen to develop. Notice that a significant tail still exists at the rump end.

Horizons 17-18

By this time, days thirty four to thirty seven, the embryo is nearly two-thirds of an inch long. The head gets still larger and the finer development of the face becomes noticeable. For example, the eyelids, the ear lobes, and the tip of the nose are now distinguishable. Internally the male gonads are clearly differentiated.

Horizons 19-21

After reaching nearly an inch by the end of the thirty-eighth to forty-third day, the trunk starts to straighten out, fingers and toes are now clearly visible at the extensions of the arm and leg buds, and most of the internal organs are formed.

261

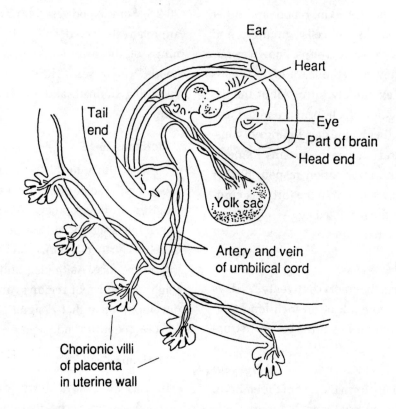

Ear

Heart

Tail end

Eye

Part of brain

Head end

Yolk sac

Artery and vein of umbilical cord

Chorionic villi of placenta in uterine wall

Figure 20 - 2 Development of the human embryo at approximately 20 days, when it is still less than one-fifth an inch long.

Horizons 22-23

The embryo, now one and one fourth inches long by days forty-four to forty-eight, has an erect head. Its limbs are long, the ears are clearly formed, and the eyelids are distinct. By the fifty-sixth day, the embryo is about one and one-half inches long and has a definite human appearance. The head is bulging and round with a high forehead. The "tail" that was present has disappeared and a rump can be seen. At this point in time the embryo has about completed its development and the fetus begins to grow.

Development of the Fetus

During this period, from about the third month to birth, there is: (a) much growth in general, (b) a finer development of the organs, and (c) a reshaping of the parts of the body until each part develops into its proper

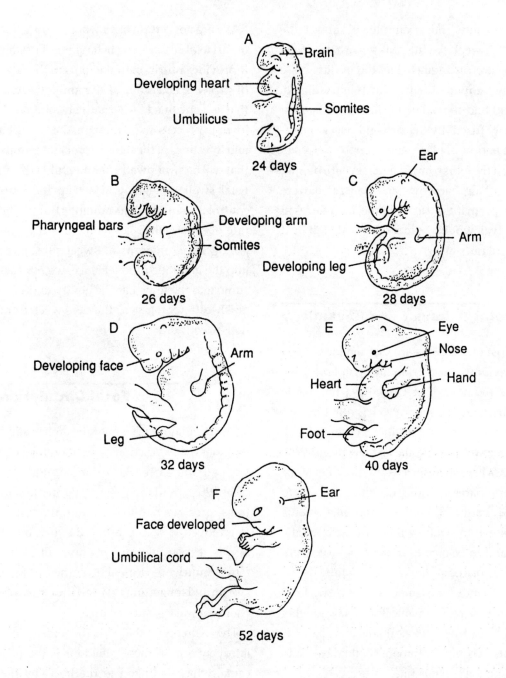

Figure 20 - 3 Human embryos between 3 1/2 and 4 1/2 weeks of age.

proportions. For example, by about the twelfth week the internal sex organs of the fetus are distinguishable, the eyelids close (they reopen during the twenty-eighth week) and the palate of the mouth is completely fused. By the sixteenth week, heartbeat and muscular activity can be detected. When the fetus completes its seventh month, it has hair, eyebrows and nails. If it were born prematurely at that time, its chances of survival would be good provided it had careful nursing.

Fetal Behavior and Physiology

Most of us are aware that the fetus has a behavior pattern from the palpable little kicks and pokes that it executes while inside the mother. The fetus also has other behavioral patterns, patterns that we can not feel. For example during the last few months the fetus is known to suck its thumb, a behavior that prepares it for the suckling it must do to survive after it is born. In addition, the fetus is known to swallow fluids of the amniotic sac and to be able to respond to sweet substances placed in the amniotic fluids. By the ninth month, for example, the fetus swallows about a quart of amniotic fluid a day, roughly one-third of the amniotic fluid there. Likewise, it puts an equivalent amount of fluid back into the amnion as fetal urine.

We are not certain as to what the function of this fetal swallowing behavior is. Possibly it provides nutrition to the fetus in the form of proteins taken out of the amniotic fluid that is consumed. Alternatively, the swallowing process may prepare the stomach and intestines of the fetus to process the milk that will be swallowed after the child is born. Fetal swallowing may also help the fetus control the volume of its amniotic fluids. The rate of swallowing can be increased by injecting small amounts of sweet substances into the amniotic fluids. Hence, we may assume that the fetus has the ability to distinguish different tastes, that is sweet from non-sweet.

Fetal Circulation

The circulatory system of the embryo and fetus has many functions, and, therefore, develops quite early. For example, not only does the circulatory system function to transport oxygen to all parts of the body and to remove the waste carbon dioxide, but it also is responsible for obtaining all of the fetus's nutrients from the mother's blood stream and for getting rid of all the embryo's chemical wastes, such as urea.

The main organ that aids the fetus' circulatory system in these functions is the placenta, which is linked to the heart by the blood vessels in the umbilical cord. The

A

Circulation in fetus (the blood contains varying amounts of both O_2 and CO_2)

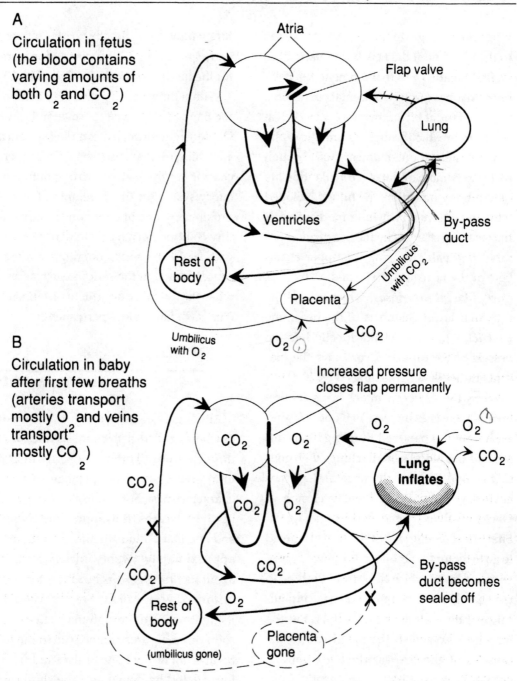

B

Circulation in baby after first few breaths (arteries transport mostly O_2 and veins transport mostly CO_2)

Figure 20 - 4 Comparison of circulation in fetus and newborn child.

lungs do not begin to function to get oxygen to the blood until the new born infant takes its first breath. Until that moment, the new-born breathes through placental circulation.

The infant at birth, therefore, has the capacity to breathe through its umbilical-placental system at one moment, and through its lungs a moment later. To make this life determining transition, the fully developed fetus must have a switching mechanism. It makes these changes by the use of a temporary *"flap valve"* linking the upper chambers of the heart, the *atria,* and a "by-pass duct" (*ductus arteriosus*) which allows the muscular lower chambers of the heart, the *ventricles,* to pump blood directly into the vessels of the umbilical cord and the placenta rather than to the lungs (Fig. 20-4).

When the newborn infant takes its first breath, there is an increase of bloodflow into the lung as it expands while filling itself with air, and this blood from the lungs then goes to the upper heart chamber and pushes against the flap valve, thereby closing it. Once the valve is closed, two things happen. First blood on one side of the heart is forced to go to the lungs. Secondly, the new oxygen-enriched blood which came from the lung to the heart starts to be pumped to the umbilical cord through the by-pass duct. But the increased oxygen in the blood causes the muscles of the by-pass duct to contract, eventually closing it off permanently. Consequently the oxygenated blood from the

lungs now goes directly to the rest of the body and normal lung circulation persists for the lifetime of the individual (Fig. 20-4).

Within time, usually a number of months, the flap valve becomes permanently sealed. Obviously with such a complex system and switching mechanisms, there are numerous chances for things to go wrong in the circulatory system of the developing fetus. Approximately one of every one hundred and fifty children born have circulatory systems with slight deviances ranging from the inconsequential to the serious, such as occurs when the by-pass duct (ductus arteriosus) or flap valve fail to close permanently.

Full-Term Fetus

Although we all like to think that our own newborn, called a *neonate,* is special and different from all other newborns, most full term fetuses have a certain range of common characteristics. For example, the average newborn weighs from approximately seven and one-half pounds to nine pounds, and is approximately twenty inches long. For about its first week, its head is often a little distorted because of the pressure caused by its tight passage through the birth canal. A soft *fontanelle* area can be seen in the head because all of the bone of the skull has not formed and hardened yet. The cheeks usually bulge because there are pads of fat present there which help the new born suckle more easily. It has a pot belly caused by the relatively large liver that developed during the fetal stage. Some newborns have hair on their heads that is often lost. On delivery, all are covered by a cheese-like fatty substance made from dead skin and secretions from the oil glands; this material helps protect the skin of the fetus from the amniotic fluid. Finally, newborns usually have blue eyes and are relatively pale; other eye colors usually develop by the time the child is six months old. Darker skin colors develop the longer the child is exposed to light.

Chapter 21

Development of the Organs of the Reproductive System

As I have stated in the last two chapters, the complete development of any organ or system in the embryo and fetus is quite complex, and it would be a futile exercise for non-biologists to study the development of all of them in detail. The reproductive system is the only one that we will study in depth.

There are three reasons for us to focus on the embryonic and fetal development of the reproductive system: First, we already have knowledge of the anatomy and physiology of the fully developed reproductive system. Second, this example shows an interesting interplay between genetics, development, and endocrinology, and to a degree, integrates many of the basic principles of human biology which are covered in this book.

Third, there now have been a number of vivid and unusual discoveries that help make the mechanism of the development of the reproductive system easier to comprehend. For example, in central America over a dozen girls have been found who developed into males at puberty. Another case in point has been the sudden surge in some young women of a dangerous vaginal cancer previously thought to occur mostly in older women. From the material which we have studied in this book thus far, we are able to understand the basic biological mechanisms which explain both these anomalous findings.

Early Differentiation of Gonads and the Sex of the Individual

The testes and ovaries originate from two groups of precursor germ cells which, during the second and third week of embryogenesis, give rise to structures called the *germinal ridges* (Fig. 21-1). These develop during the third week into a pair of *indifferent gonads* which can not be distinguished as either male or female. By the end of the fourth week of development, however, the indifferent gonads are committed to becoming either testes or ovaries by a series of genetic and chemical events described below.

Development of Male and Female Tubules

By the fifth and sixth weeks of development, the embryo, in addition to having formed the primitive gonads, has also formed a series of three ducts (tubules) that are associated with the gonads. At first the embryo has sets of both female and male ducts (Fig. 21-2A): The ones that end with a swelling close to the primitive gonads are called *Wolffian ducts*. These persist in embryos that become males and develop into the epididymis and vas deferens (Fig.21-2A,B). The other set, called *Mullerian ducts*, persist in embryos that become females, and develop into the fimbria, oviducts, uterus, and vagina (Fig. 21-2A,C). The complex processes that cause one set of tubes to persist (Wolffian in males, Mullerian in females), and the other to degenerate and disappear (Mullerian in males, Wolffian in females) all seem to be triggered by the presence or absence of the Y chromosome in the embryo.

Sequence of Events Leading to the Development of the Internal Genitalia of Males

As pointed out earlier, maleness is determined at conception by the presence of a single Y chromosome in the fertilized egg. Even individuals having two X chromo-

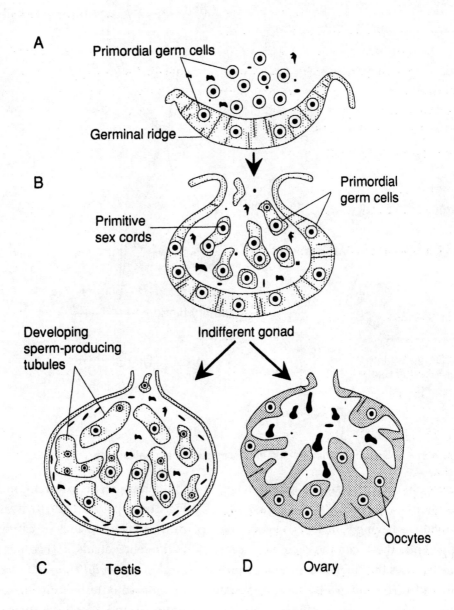

A

Primordial germ cells

Germinal ridge

B

Primordial germ cells

Primitive sex cords

Indifferent gonad

Developing sperm-producing tubules

Oocytes

C   Testis

D   Ovary

Figure 21 - 1 Development of early gonads in embryo

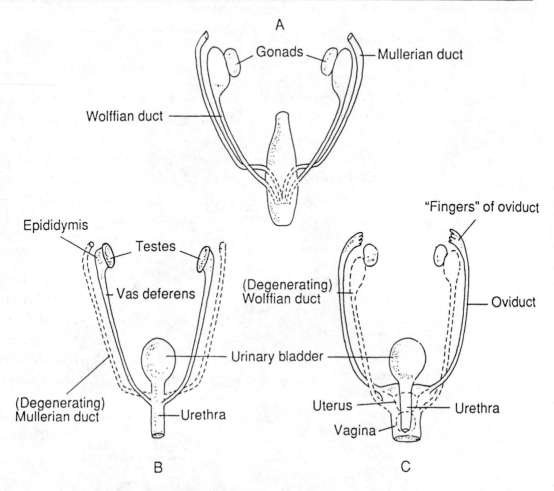

Figure 21 - 2 Transformations of the genital ducts in mammalian embryos.

somes can become males as long as one Y chromosome is present as is the case with individuals having Klinefelter's syndrome. Apparently the Y chromosome has a gene which causes the embryo to produce a protein, called **H-Y antigen** by some, that sets into motion a series of events which cause that embryo to develop male characteristics.

This protein is produced by the XY embryo early in development, and it directs the indifferent gonads to develop into testes (see Fig. 21-4, top line). To do so, it causes the sex cords, which are coils of tissue within the indifferent gonads, to become the **sperm-producing (seminiferous) tubules.** Soon after the embryonic testes form, the testes start to produce and secrete a substance called **MDIF**, an abbreviation for **Mullerian duct inhibiting factor**, and testosterone (Fig. 21-4, second line). From its name, obviously the

MDIF stops the Mullerian ducts from developing any further and causes them to degenerate and disappear. Simultaneously the embryonic testes begin to secrete testosterone; this hormone stimulates the Wolffian ducts to develop into the epididymis and vas deferens of the male genitalia. These changes all take place during the embryonic stage of development. The sequence of these and the events described on pages 274-279 can be followed more easily by referring to the flow chart in Figure 21-4.

Development of the External Genitalia

The external genitalia form during the fetal stage (Fig. 21-3). As in the case of the various ducts and tubules which become part of the internal genitalia, the embryo and fetus also form a number of developmental precursors to the external genitalia. Those precursors can become either male or female genitalia depending upon the hormones being secreted at the time.

There are three key precursors: one is the genital bulge [tubercle], which can become the clitoris in females or the head of the penis in males. Second, there are the urogenital folds which, in females, become the labia minor, which is the opening to the vestibule

Table 21-1 Fate of Embryonic Precursors to Genitalia		
Structure in Early Embryo	**Structure in the Male**	**Structure in the Female**
Indifferent gonad	Testis	Ovary
Primary sex cords of gonad	Sperm-forming tubules	Disappears
Wolffian duct	Epididymis and vas deferens	Disappears
Mullerian duct	Disappears	Oviduct, uterus
Genital bulge	Head of penis	Clitoris
Urogenital folds	Raphe and body of penis	Labia minora
Genital swellings	Scrotum	Labia majora

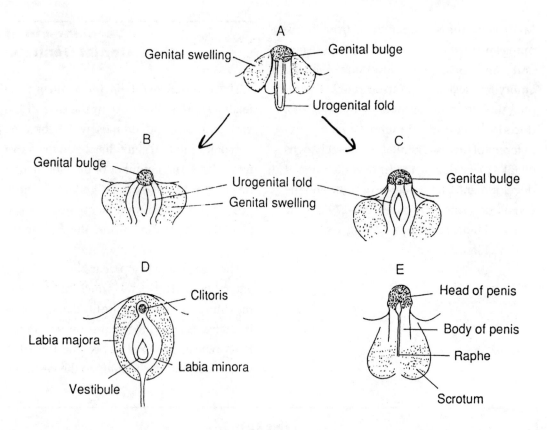

Figure 21 - 3 Development of the external female and male genitalia.

and vagina. In males, the urogenital folds round up and fuse to form the main body of the penis. The site of this fusion remains as a linear "scar" on the underside of the penis called a *raphe*. Lastly, there are the *genital swellings*, first seen in the early embryo. In females these become the labia majora, whereas in males these become the scrotum. The potential fates of the embryonic precursors to the genitalia are listed in Table 21-1.

Sequences of Events Leading to the Development of External Genitalia in Males

Referring to the flow chart (Fig. 21- 4), we can see that the presence of the testes is also the key factor in the setting into motion the events leading to the development of the external genitalia in the fetal stage. During this stage, the testes become very rich in an enzyme that converts the male hormone, tes-

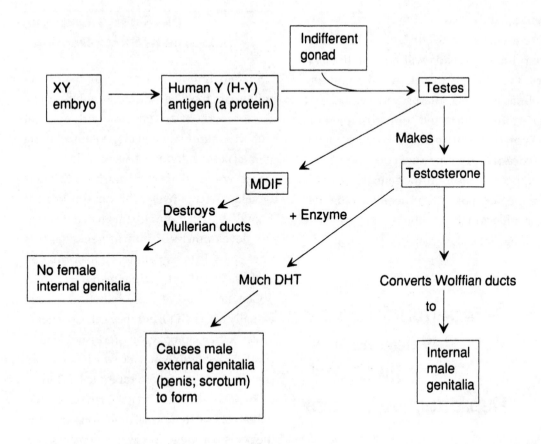

Figure 21 - 4 Genetic and hormonal events leading to development of male genitalia.

tosterone, into a slightly modified form called *DHT*, an abbreviation for *dihydrotestosterone*. The DHT is the critical hormone that stimulates the genital bulge to become the head of penis, the urogenital folds to become the body and raphe of the penis, and the genital swellings to become the scrotum. Most of these events occur by the sixth month. By the seventh month the testes descend into the scrotum.

Development of the Internal and External Female Genitalia

Compared to the events outlined in the flow chart (Fig. 21-4) for the genetic and hormonal control of the development of the male genitalia, the events controlling the development of the female genitalia appear quite simple. Females possess no Y chromosome and hence produce no H-Y antigen.

Consequently there will be no testes to produce testosterone, MDIF, and DHT. Hence, the primitive gonads will become the ovaries, the Mullerian ducts will become the fimbria, oviducts, uterus and vagina, and the Wolffian ducts and sex cords, for some unknown reason, perhaps because of the absence of testosterone, will degenerate and disappear. In addition, the genital bulge will become the clitoris, the urogenital folds - the labia minora of the vestibule, and the genital swellings - the labia majora.

Aberrancies in the Development of the Sexual Reproductive System

The chain of events leading to the development of maleness or femaleness is so complex that there exist many chances for deviations to occur. In recent years a number of unusual cases have been reported that illustrate the relative importance of each of the stages in development of the reproductive system outlined in the flow chart in Figure 21-4.

Case #1: The Gueve Doces of Santa Domingo

In two remote villages near Santa Domingo in the Dominican Republic, there were reports of sixteen little girls who, at puberty, suddenly developed into young men. These individuals were called gueve doces by the villagers, Spanish for "a penis develops at age twelve." From our present knowledge of the mechanisms controlling the development of human genitalia, it is now possible to explain this transformation of those "girls" into young men (Fig. 21-5).

Genetically all of these individuals were actually males (XY), but they carried a gene, which when homozygous recessive, would produce a defective enzyme which would not convert sufficient testosterone into DHT to do its job. Thus the fetus did not form male external genitalia at the normal time. Consequently, the penis of a gueve doce was extremely small, more like an enlarged clitoris, the body of the penis and the raphe never formed, and the testes never descended because the genital swellings remained as the labia majora rather than becoming the scrotum.

At puberty, however, the brain triggered the release of FSH and ICSH from the anterior pituitary, and the interstitial cells of the undescended testes started to produce large amounts of testosterone. With such an abundance of testosterone, even the small amount of testicular enzyme present in the

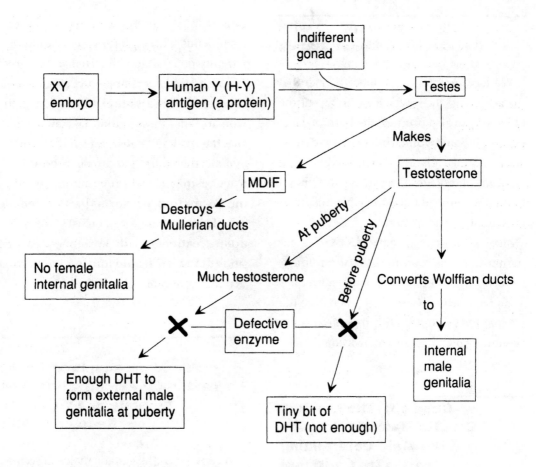

Figure 21 - 5 Genetic and hormonal events leading to development of genitalia in Gueve Doces individuals.

testes of a gueve doce was able to make sufficient DHT so that to various degrees external male genitalia finally developed. Thus, with this burst of testosterone occurring at puberty, the DHT formed from it caused the clitoris to become a small but functional penis, and the testes to descend into the newly formed scrotum. In addition, the increased testosterone formed by the testes caused these youngsters to take on the secondary sex characteristics of males, such as a lower pitched voice and facial hair.

A study of these sixteen individuals showed that at least one of the young men became a well developed male. Others, though less developed, were able to have erections and sexual intercourse. Of these sixteen individuals, fourteen identified as males after puberty, and two still considered themselves females.

Case #2: Females with Overactive Adrenal Glands

We have not yet mentioned the fact that the adrenal glands, which sit on top of the kidneys, can also produce the hormone testosterone. Some women have enlarged overactive adrenal glands which secrete large amounts of testosterone, causing the clitoris to become enlarged and the labia majora to form a sac like a scrotum.

Most such genetic females (XX) are extreme "tomboys" as young girls. A number have been treated with female hormones and given corrective surgery to enlarge the vagina, for example. With treatment some eventually ovulated and menstruated.

Case #3: The Inability of Genetic Males to Produce the Male Determining Protein (H-Y antigen)

Some genetic males (XY) have a mutation which does not allow them to produce the H-Y antigen. As a consequence, the primitive gonads do not become testes and no testosterone and MDIF are produced. With no MDIF present, the Mullerian ducts remain and form the female internal genitalia including a good uterus. Their primitive gonads do not develop a perfect ovary, possibly because of some genetic interference caused by the presence of the Y chromosome. Thus, although these individuals apparently possess good internal and external female genitalia, the imperfect ovary makes such an XY individual into an infertile woman. Since these individuals can not ovulate, they can not menstruate. If they can not ovulate, then the ovary can not produce sufficient estrogen and progesterone to aid in the formation of normal-sized female breasts; they can form breasts, however, upon treatment with hormones. In one study of four such individuals, two married as females, and one adopted children.

Case #4: Pseudohermaphrodites by Testicular Feminization: The Inability of XY Cells to Recognize DHT

In contrast to the genetic males described in Case #3, genetic males (XY) who become *pseudohermaphrodites* can produce the H-Y antigen. Thus, as shown in Figure 21-6, pseudohermaphrodites develop testes which then produce MDIF and testosterone. The MDIF, as one would expect, destroys the Mullerian ducts, and no internal female genitalia develop. The testosterone also carries out its normal function of stimulating the Wolffian ducts to develop into the vas deferens and epididymis.

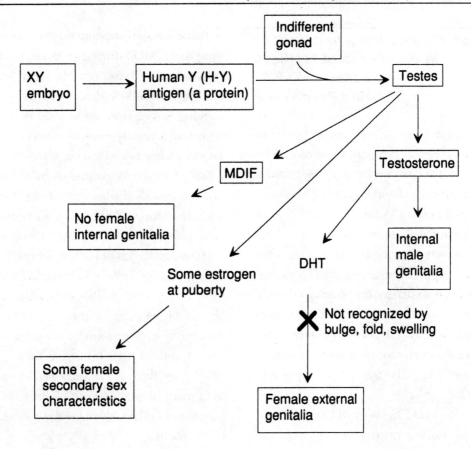

Figure 21 - 6 Genetic and hormonal events leading to development of pseudohermaphrodites.

There is, however, a problem with these XY individuals: the cells of the genital bulge, fold, and swellings do not recognize and respond to the DHT produced from the testosterone. Consequently, the tissues that are precursors to the external genitalia develop as would those of a female, and not as those of a male. Thus, these individuals have internal male genitalia and external female ones.

This imbalance of the internal and external genitalia apparently affects the production of hormones by the testes at puberty, be-

cause at that time they seem to produce more estrogen than they would in normal XY males. Thus, the extra estrogen secreted as a result of this testicular feminization stimulates the pseudohermaphrodites to take on some of the secondary sex characteristics of females.

Case #5: Infertility and Cancer in Daughters of Mothers Exposed to D.E.S. During Pregnancy

Of all the birth defects caused by drugs taken by pregnant mothers, those caused by D.E.S. are the most unusual. Why? Because most drug-related birth defects can be detected in the infant within a few months after birth. With D.E.S., to the contrary, the deleterious effects are not detected until the child has passed puberty and is ready to conceive children. That is to say, the effects caused by the D.E.S. that the mother took remain dormant in the child for years, to become activated only later by the high levels of steroid hormones released in the body from puberty onwards.

But what is D.E.S.? Why did women take the drug during pregnancy? D.E.S. is the abbreviation for *diethylstilbestrol*, a powerful and inexpensive synthetic estrogen often given to cattle and chickens to make them fatter. About thirty years ago D.E.S. was widely prescribed to pregnant women on the premise that it would help prevent miscarriages although there was no concrete evidence to support such a claim. It is now believed that there are over a million young females in the United States today who are deleteriously affected by the D.E.S. taken by their mothers.

The effects on the daughters of mothers who took D.E.S. during pregnancy can be quite serious. For example, between 1971-78 doctors reported two hundred cases of girls ranging in age from eight to twenty eight who had a deadly vaginal cancer (*vaginal adenocarcinoma*) which up until then was usually found only in older women. Twenty of these girls died from that cancer. Medical scientists estimate that one out of every thousand girls born to mothers who took D.E.S. will develop vaginal adenocarcinoma.

There is another harmful effect of D.E.S. on those daughters, one that will not kill them but which can render them infertile. This condition, which produces changes in the surface cells of the vagina and cervix, may affect from thirty to seventy percent of the daughters, depending upon the week of pregnancy during which the drug was taken by the mother.

Because such a large proportion of women exhibiting these syndromes were daughters of mothers who took D.E.S., less than eight weeks into pregnancy, it is believed that the hormone interfered with the normal development of the Mullerian ducts into the vagina and cervix. Usually these structures, along with the uterus, develop from the fusion of the bottom portions of two Mullerian ducts. Apparently the high levels of D.E.S. affect this fusion so that the cells of the surface of the vagina will respond adversely to further doses of estrogen, such as are released in puberty.

Even though we still do not have answers to some of these questions about the actions of D.E.S., this example highlights some serious points and warnings. For one, doctors and pregnant mothers have a great responsibility to the child. Thus, much research must be done on drugs for pregnant women before they are allowed to be prescribed. The tragedy of D.E.S. is that ten years before D.E.S. was given to pregnant mothers, biologists had published similar findings observed in mice exposed to D.E.S.

Secondly, because steroid hormones exacerbate the effects of D.E.S. in daughters, those individuals should not take steroid contraceptive pills, nor should they use steroids for therapy during menopause. Possibly an increase in steroid levels might enhance the chance for the deadly vaginal adenocarcinoma to develop.

Lastly, although we have focused on daughters of D.E.S. mothers, the sons also appear to be in some danger. Earlier research on male mice whose mother had been given D.E.S. showed that some had problems forming sperm, and had an increased proportion of epididymal cysts. Newer evidence shows that in humans there are cases of young males whose reproductive systems are damaged because of D.E.S. taken by their mothers. To date, no disproportionate numbers of cancers have been detected in sons of mothers who have taken D.E.S. during pregnancy.

Chapter 22

Placental Physiology

The embryo and fetus have a continual need to obtain food and oxygen, and to get rid of food wastes and carbon dioxide. To carry out such exchanges, mammals have evolved a rather unique temporary organ - the *placenta*. This organ, which develops in the wall of the uterus during pregnancy, is linked to the developing child by means of the *umbilical cord*. At the place where the placenta imbeds into the uterus, the blood vessels of the embryo and fetus come very close to those of the mother.

This close juxtaposition of blood vessels allows food, wastes, and gases are exchanged between the fetus and mother. Food wastes, such as urea and carbon dioxide, for example, are taken up from the blood vessels of the placenta by the mother's blood vessels and are eliminated by means of her kidneys, bladder, and lungs. Because the placenta is the organ through which the developing child and the mother exchange materials, it is also the place where drugs and other substances present in the mother enter the blood vessels of the embryo and fetus.

The placenta also serves as a site for the manufacture of the hormones human chorionic gonadotrophin (hCG), estrogen, and progesterone. Hence, we might also call the placenta a temporary endocrine gland.

Shape and Composition

The shape of the placenta varies from mammal to mammal. In humans, it is shaped like a round hollow cake about nine inches in diameter and one inch thick all around. The word placenta, in fact, comes form the Latin meaning "formed like a cake." The average placenta weighs about a pound.

The fetal tissue making up the placenta comes mostly from the various layers of the trophoblast and contains some of the membranes of the amnion and yolk sac. The blood vessels that run throughout the placenta link up with the artery and vein of the umbilical cord.

The placenta is smooth on the side facing the cavity of the uterus; the other side of the placenta, the "rough" side, faces the uterine wall and is covered with thousands of finger-like projections called villi.

The *villi* contain the small blood vessels which are juxtaposed with the small blood vessels of the mother's endometrium. Blood vessels are seen in the villi as early as the third week of embryonic development (Fig.22-1). On these villi are still smaller projections, called *microvilli*. Even though the villi and microvilli are surrounded by maternal fluids in the endometrium, the blood in the fetal and maternal blood vessels usually does not mix.

The thousands of little projections coming out of the placenta's villi and microvilli greatly increase the surface area of the placenta contacting the endometrium. Hence, there is a greater chance for a more rapid and extensive exchange of food and wastes between the mother and developing child. Even though the placenta is about nine inches in diameter, the villi and microvilli help to increase its surface area so that it makes contact with the endometrium over a space of a dozen square yards.

Transfer of Materials through the Placenta

Substances other than gases (oxygen and carbon dioxide), food substances (glucose, amino acids, fats, and vitamins) and wastes (urea) can pass through the placenta. These substances may include certain drugs, alcohol, carbon monoxide, and even the anaesthetics that some mothers take before and during delivery. Such substances can cause many of the birth defects described in Chapter 23.

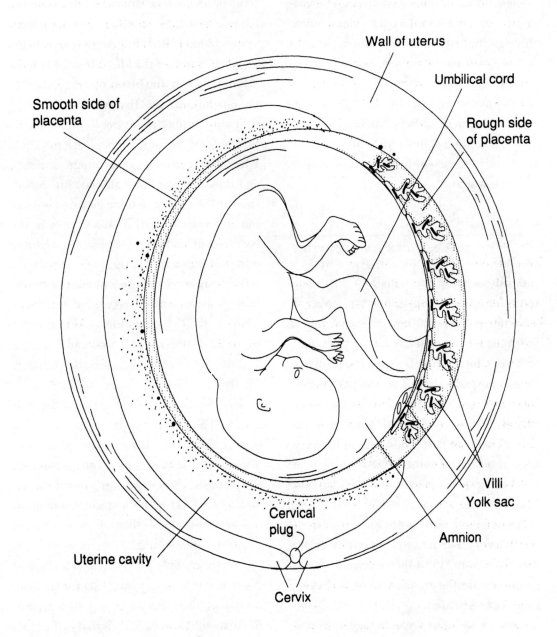

Figure 22 - 1 View of embryo, placenta, villi, and amnion in the uterus.

Most substances that pass through the placenta are small molecules which enter through the cellular membranes of the cells in the small blood vessels. Some of these substances, such as oxygen and certain fat soluble materials, can pass through the placenta unaltered. Others, such as sugars, amino acids and soluble iron salts, must be modified in some way by the cell before they can be transferred.

Placenta as a Barrier

Besides serving the fetus by exchanging foods for wastes, the placenta can also act as a barrier to keep out materials that might harm the fetus. Relatively large objects and materials, for example bacteria and proteins, do not normally cross into the placenta. If bacteria could cross this barrier, the fetus would be susceptible to all sorts of infections. There are cases, however, where the toxic products of some bacteria present in the mother do cross over into the placental blood vessels.

A number of viruses are known to cross the placental barrier, and they may prove extremely harmful to the embryo or fetus. Examples are the measles virus and cytomegalovirus (CMV).

Even the slightest break in the blood vessels of the placenta and uterus may allow some proteins to pass from the mother to the fetus or vica versa. Consider the case of an Rh negative (Rh⁻) mother carrying a fetus whose blood is Rh^+. If there is a break in the placenta, some of the blood from that fetus might pass into the blood of the mother. In the ensuing months that mother will build up many antibodies to the Rh^+ blood cells that entered her circulatory system. Thus, when that mother becomes pregnant again, and if that second child also has Rh^+ blood, there is the chance that the elevated level of the antibody to the Rh^+ blood cells in the mother will lead across the placental barrier into the fetus and destroy its blood cells.

It is important to note that the Rh⁻ condition is a autosomal recessive trait (rr), whereas an Rh^+ individual could be either RR or Rr. Hence there is a potential danger to the second fetus only when the mother is Rh⁻ (rr) and the father is Rh^+ (RR or Rr) (Fig. 22-2). If the father is RR, then all of the children will be heterozygous for Rh^+. In such a case, the first baby will usually survive; but if some of that baby's blood seeps through the placenta into the mother's blood stream and antibodies to the Rh^+ factor build up in the mother's blood, then all of her subsequent children would be in danger.

If, on the other hand, the mother is Rh⁻ and the father is heterozygous (Rr) for the Rh^+ condition, then there is a fifty-fifty chance the first child will be Rh^+. If that is the case, then all subsequent Rh^+ children will be in danger. If the first child is Rh⁻, however, no

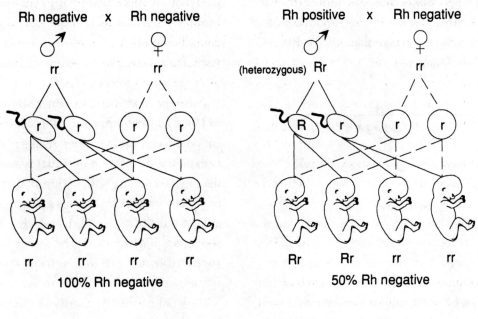

Rh negative x Rh negative

♂ ♀

rr rr

rr   rr   rr   rr

100% Rh negative

Rh positive x Rh negative

(heterozygous) ♂ ♀

Rr rr

Rr   Rr   rr   rr

50% Rh negative

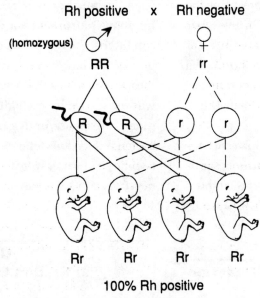

Rh positive x Rh negative

(homozygous) ♂ ♀

RR rr

Rr   Rr   Rr   Rr

100% Rh positive

Figure 22 - 2 Possible genotypes of children born to Rh negative (Rh⁻) mothers.

anti - Rh$^+$ antibodies will build up in the blood stream of the mother, and the second child, whether it be either Rh$^+$ or Rh$^-$, will not be endangered.

Passage of Hormones

We have already discussed the role played by the placenta in making and secreting the pregnancy hormone, hCG, and the steroids estrogen and progesterone. The placenta also serves two other functions in relation to hormones. For one, it can act as a barrier preventing normal levels of maternal hormones from entering the embryo and influencing changes in the development of the embryonic and fetal organs. On the other hand, it can not prevent extra large doses of some synthetic and powerful hormones present in the mother's circulation from entering the embryo as was the case with D.E.S. (Chapt. 21).

The second function of the placenta related to hormones is to carry hormones produced by the fetus to the wall of the uterus in order to initiate the onset of labor (Chapt. 25).

Decidual Reaction

It may seem strange that such an important and versatile organ as the placenta, which becomes mature three months into gestation, is discarded by the body just six months later. But such is the case, the placenta being shed at term once the child is born. For that reason it is given the ignominious name of "afterbirth."

This separation process, which starts early in labor as the uterus begins to contract, is called the *decidual reaction*. As the placenta is expelled during the decidual reaction, it usually takes along with it some of the tissue of the uterine wall and leaves a raw bleeding area at the site in the uterus which the placenta had just occupied. The uterus is very susceptible to infection during the few weeks needed for that site to repair itself.

The chances of the uterus becoming infected are much greater if the birth is assisted by a physician or midwife under non-sterile conditions. Such was the case in the latter part of the nineteenth century when, with the increase of hospital assisted births, there was an epidemic of childbed fever in Europe. In response to this high incidence of bacterial infections following deliveries, antiseptic procedures were introduced into the maternity ward and into hospitals in general.

Use of Villi to Predict Birth Defects

In recent years, medical scientists have used pieces of chorionic villi excised from the placenta to analyze for chromosomal aberrancies and other types of birth defects. Up until recently, about the only way to obtain embryonic cells for diagnosis was to use the very delicate and expensive process of amniocentesis. New research is showing that similar diagnoses can be carried out using embryonic cells taken from the chorionic villi of the placenta. During this procedure, the physician removes a little plug of the villi from the wall of the uterus without touching the amniotic sac. The embryonic and fetal cells are removed from the plug and are analyzed for karyotype or other factors just as is done with embryonic cells obtained by amniocentesis.

Scientists still do not know for sure if removing a small plug of placental villi affects the development of the embryo detrimentally, although to date it appears not to do so. There are high hopes that once this technique is perfected, it will revolutionize prenatal testing and will make it available to more people at a reasonable cost.

Chorionic villi sampling (CVS) is performed during the third month of pregnancy whereas amniocentesis is usually performed between the 15th and 20th week of pregnancy or a little earlier. Because CVS can be carried out earlier, it does reduce anxiety in the patient and decreases the risk when a birth defect is detected and an abortion is carried out. On the other hand, CVS can lead to miscarriages, and it is not as effective as amniocentesis in determining certain kinds of birth defects, such as anencephaly (absence of a brain) and spina bifida (Chapt. 23).

Chapter 23

Environmentally Induced Birth Defects

When a child is born with an abnormality, whether it be physical, such as a malformed organ or limb, or mental, as in mental retardation, we speak of these abnormalities as "birth defects." These defects may fall into three broad categories: First, chromosomal aberrations, which usually result from a faulty cell division leading to the formation of an abnormal egg or sperm, or from an error in fertilization. These can produce such syndromes as Down's (trisomy 21), Klinefelter's (XXY) and Turner's (XO) (Chapt. 12).

A second category consists of genetic diseases which result from mutations (permanent chemical changes in genes) that are inherited from one generation to the next and are usually expressed in individuals homozygous recessive for that trait. Examples are sickle cell anemia, hemophilia and Tay-Sachs disease (Chapts. 14-16).

This chapter, however, deals with a third broad category that accounts for approximately eighty percent of all birth defects and affects to some degree from two to five percent of all children born. I am referring to those birth defects caused by environmental agents which originate from outside of the embryo. A number of *environmental agents* do little harm, if any, to the pregnant woman, but can cause abnormalities to develop in the embryo and fetus. Five general classes of those agents, called *teratogens*, are: (1) *Drugs* (medicine, antibiotics, hormones, social drugs); (2) *Infectious agents* (viruses, venereal diseases, parasites); (3) *Pollutants* (in food, air, water, and the work place); (4) *Radiation* (e.g., X-rays); and (5) *Extremes of diet* (malnutrition, excess of vitamins).

Except for radiation, all of these teratogens enter the embryo and fetus by being transferred from the mother's bloodstream to the embryo's bloodstream by means of the placenta. Hence, a major feature of most teratogens is that they must pass through the placental barrier. Unfortunately the list of known deleterious materials and agents which enter the mother's body and that can pass through the placenta is ever increasing.

Time During Pregnancy that Teratogens Act

Not all teratogens are harmful during all stages of pregnancy. Let us consider each of the three major stages. During the stage from the fertilized egg to the blastocyst, i.e. during the first twelve days of pregnancy, most teratogens do little harm. If any injury is done, however, the defective blastocyst usually is not able to implant and is aborted unnoticed by the mother.

Insofar as susceptibility to teratogens is concerned, the most critical period of gestation is the embryonic period (middle of the first through the third month); during this stage the organs are formed and the embryo takes on the characteristics of a miniature human being. Most of the best known teratogens, whether they be drug, radiation or infectious agent, do the most damage during this period.

Unfortunately, most women do not even know that they are pregnant until three to six weeks into pregnancy. By the time a pregnant woman has missed her first menstrual period, the embryo has started to form such organs as the heart, brain and spinal cord. Other major organs begin to form soon afterwards, usually before most women have their first appointment with a physician to confirm their suspected pregnancy. If during that time the women have had X-rays, or have been taking teratogenic compounds which may be harmless to individuals who are non-pregnant, then there is the possibility that the embryo may have already undergone serious and irreparable damage.

About the only sensible advice to give women who are trying to become pregnant is to avoid all sources of teratogens. Likewise, *potential* mothers, i.e. any ovulating women engaging in intercourse, should also avoid exposure to teratogens.

During the *fetal period*, chances are reduced of most teratogens inducing serious birth defects because the organs have already taken their basic form. Nonetheless, teratogens may cause fetal organs to develop in a weakened condition such that the health of the child will suffer later in life. More recently medical scientists find that even the anaesthetics given to some mothers to ease the pain of birth can lead to behavioral abnormalities in the newborn, such as retarding the time it takes for the child to sit erect.

Thus the idea that the embryo and fetus are sheltered from damage in the protective waters of the amniotic sac in the interior of the mother appears to be a myth. We now know that many agents which can be tolerated by adults, pregnant or not, can cause great harm to the embryo and some harm to the fetus.

Drugs and Other Chemicals

During pregnancy women undergo a large number of physiological, physical and psychological changes ranging from nausea and constipation to mental stress (Chapt. 24). Some can cope with most of those changes, but the majority feel that they need the help of medication to lessen the discomforts of pregnancy. Thus, pregnant women may be tempted to take a number of common drugs such as laxatives, aspirins, and tranquilizers.

Most drugs, however, function by altering the workings of the body. Alterations that seem inconsequential and even helpful to the mother's body may have drastic effects on the developing embryo and fetus. For example, laxatives act by affecting body water, and thereby could alter the water balance of the embryo. Heavy doses of aspirin can be dangerous because they affect the synthesis of prostaglandins. A tranquilizer such as thalidomide can lead to children having no arms or legs.

Pregnant women who smoke cigarettes or consume what they consider to be small amounts of alcohol can damage their developing child. For example, if the mother smokes even a single cigarette, the nicotine, carbon monoxide, and carbon dioxide from the inhaled smoke eventually pass into the fetus to increase its heart beat and to cause its blood vessels to contract. Mothers who are heavy smokers are more prone to abortions and to bearing premature children; furthermore, those babies are subject to convulsions and fits more often than are babies of nonsmoking mothers.

Working Rule Regarding Drugs

In general each pregnant woman and her physician need to weigh the consequences of either taking drugs or refraining from taking them. Questions that they must ask are: Is the drug being considered the safest one for the ailment? Is the pain or inconvenience bearable so that the drug can be avoided? Is the ailment so severe and debilitating that the drug must be taken regardless of the risks?

These are not easy questions to answer. Perhaps the working rule should be: *If the need is great enough, then the risk may be worth it.* Unfortunately, however, pregnant women are not always informed of the severe consequences of taking drugs. When they are uncomfortable during pregnancy, the tendency may be to exaggerate the need and ignore the risk. Most physicians would urge that pregnant mothers avoid taking drugs, especially during the first three months of gestation.

Thalidomide

Perhaps one of the most dramatic and tragic cases of a drug causing birth defects is the experience with thalidomide. This drug was developed in Germany in the 1950s as a sleeping pill that had no "hangover." It was assumed to be harmless to pregnant women and was prescribed for them on the premise that it would also reduce their nausea. Within two years after thalidomide was introduced into the market a number of children were born with a rare birth defect, *phocomelia*, which means "limbs like a seal." Such children would lack all or parts of their limbs. Some would have heart problems, obstructed intestines, or abnormally developed ears, but their brains were not affected.

The link between thalidomide and phocomelia was finally recognized when in one clinic in Germany a hundred and fifty-four cases of phocomelia were reported within two months; all of the mothers had taken thalidomide early in their pregnancies. In the early 1960s, 6,000 cases were reported, with five thousand of them being in West Germany. Shortly thereafter the drug was called off the market. It never was approved for use in the United States, although some women got it from Germany illegally.

This unfortunate and unintended "human experiment" showed that small amounts of teratogenic drugs taken during the embryonic period caused devastating birth defects in the child. Many cases were caused by the pregnant mother taking as few as one to two tablets! The thirty-fourth to fiftieth day of gestation was the most sensitive time for thalidomide damage. This time span is usually defined in terms of days since the last menstrual period (LMP) — in other words during the third and fourth week of the life of the embryo. Ear damage appeared if the drug was taken between the thirty-four to thirty-eight days LMP; missing arms, between thirty-nine to forty-four days LMP; missing legs, between forty-four to forty-eight days LMP; and abnormal intestines and heart, between forty to forty-five days LMP.

Although thalidomide is now off the market and is no longer a threat, a huge number of other drugs are being introduced yearly. Unless proper testing has been done on mammals and approved by the FDA (Federal Drug Administration), those drugs should be avoided by pregnant mothers.

Some Life-Saving Medicines and Other Chemicals Known to be Teratogens

From the foregoing discussion on thalidomide, I do not wish to leave the impression that it is a simple matter to distinguish whether a drug is teratogenic or not, or that the choice to avoid such drugs is easy. Consider a woman who comes down with a life-threatening illness during pregnancy. For example, in order to survive, a pregnant woman who has learned she has leukemia may need to take Busalphan, a known teratogen, or other chemotherapeutic compounds of potential teratogenicity. Likewise, quinine, a drug for treating patients with malaria, if taken by a pregnant woman, can cause brain and skeletal damage in the embryo.

Caution is also advised in regard to hormones and antibiotics. For example, the hormone insulin, taken by diabetics, is known to harm the fetus in a number of ways. Even Vitamin A, when taken in excessive amounts, is teratogenic. The case is not so clear to date with antibiotics, many of which, including penicillin and streptomycin, can cause damage to the fetuses of laboratory mice. Tetracycline, another common antibiotic, is teratogenic to humans.

The list of chemical teratogens may seem endless. We have already mentioned DES (Chapt. 21) which shows its effects after the child has reached puberty. We can also add to the list such substances as many anticancer drugs, some anticonvulsants, certain male hormones, and heroin. Social drugs of abuse are still another story. Suffice it to say that not only are some of these social drugs teratogens, but many leave the newborn child with the same symptoms of addiction as are found in the addicted mother.

Pollutants

Even more subtle are cases in which pregnant mothers have no knowledge at all that they are consuming teratogens. Some of these substances may exist as environmental pollutants over which we have little control until the deleterious effect on the embryo is discovered years later.

For example, at one time the port city of Minimata, Japan, reported an unusually large number of children born with brain damage and cerebral palsy. These incidences of Minimata's disease were eventually traced to fish which the mother had eaten. Apparently those fish had a high level of mercury in them that came from wastes discharged into the bay by a fertilizer manufacturer. When the fish were consumed by pregnant women, the mercury caused birth defects in the developing children.

Fetal Alcohol Syndrome

Shakespeare once wrote: "Wine provoketh desire, but taketh away the ability." Today we also know that wine and all alcoholic beverages can cause fetal alcohol syndrome, now thought to be the third leading cause of mental retardation in the United States today. Chronic alcoholism in prospective mothers has been recognized as a major cause of birth defects only in the last decade. Characteristics of infants with fetal alcohol syndrome include a characteristic "short nosed" facial expression, prenatal onset of growth deficiency, creases in the palms, a small head and brain, tremor seizures, hypertension, heart problems, and mental retardation.

Varying degrees of this syndrome affect more than one-fourth of all children of mothers who are chronic alcoholics. To date medical scientists do not know what the dangerous levels of alcohol consumption are, and obviously, it will vary with each individual and the diet and lifestyle of that individual. Nonetheless, in certain cases even moderate drinking has been shown to cause some degree of fetal alcohol syndrome in newborns. In one study, abnormalities were shown to occur in twenty percent of the children of mothers consuming about two ounces of alcohol a day, in ten percent of those consuming one to two ounces a day, and in two to five percent of those consuming an ounce a day. Even one ounce of alcohol consumed close to term has been shown to suppress fetal breathing.

A practical guideline is that an ounce of alcohol can be found in either an ounce of hard alcohol (whiskey, gin, scotch), a small glass of wine, or a bottle of beer. Until more research is done on this subject, I would recommend that the safest dose of alcohol to be consumed by pregnant mothers is NONE - not even a social toast.

Range of Physical Birth Defects Caused by Teratogens

A wide range of abnormalities is caused by teratogens ranging from such harmless but unusual defects as polydactylism (having an extra finger or toe) to birthmarks and to a host of more serious defects such as: cleft lip and cleft palate, club foot, heart defects, mental retardation, spina bifida (defects of spinal vertebrae and cord), phocomelia, pyloric stenosis (partial closure of opening between stomach and small intestine), various abnormalities of the brain such as microcephaly (small head), hydrocephaly (excess fluids in the brain), anencephaly, and ear and eye defects. I do not intend to delve into a medical discussion of each of these defects; there are ample books on the subject. The major object of this book is to give you an understanding of how these defects are caused, and therefore, how they can be avoided.

Diet

Many examples of dietary teratogens might be classified as drugs. For example, on the one hand alcohol may be considered a food with high caloric value, and on the other hand, in excess it can cause fetal alcohol syndrome. Even some vitamins, such as the fat-soluble Vitamins A and D which are required by all individuals including the developing fetus, in excessive amounts can act as teratogens.

Teratogenic effects on the embryo can also be caused by deficiencies in essential vitamins, minerals and amino acids. Hence, the pregnant woman should have a well-balanced diet with sufficient nutrients for herself and for the developing child.

One of the most serious causes of birth defects today is malnutrition. During any long period in which food consumption is restricted, the normal adult forms substances called ketone bodies. Although the liver of adults is able to handle these toxic substances, the liver of the fetus can not. As a consequence, irreversible damage to the brain usually occurs and the overall growth patterns of the child are retarded. This form of severe malnutrition and its long-term effects on children born to malnourished mothers is a tragedy that will have major social and political effects not only in countries hit by famine, but also in the world's large cities where the poor do not get enough to eat.

Malnourishment also can affect the fetus of well-to-do women who, for cosmetic reasons, want to look slimmer. It is now known that women who diet heavily during the last month of pregnancy give birth to children who have some of the same birth defects seen in children of malnourished women.

Lastly there are food additives, some unintentional as in the case of the mercury poisoning of fish in Japan, or intentional, when compounds "that may be hazardous to your health," according to the labels, such as saccharin, pesticides, and food preservatives, are added to foods made available commercially.

Infectious Agents

In Chapter 8 on venereal diseases we have already learned how syphilis and herpes can affect the unborn. Yet diseases harmful to the embryo and fetus are not only those that are sexually transmitted. There are a number of viruses, some that are relatively less harmful to the mother, which can cause serious birth defects in her child. Apparently the tissues of the embryo and fetus are more susceptible to the viruses because they do not have a well developed defense system to get rid of the viruses as do adults. Most of thee infectious agents are viruses; bacteria

that infect the mother fortunately cannot pass through the placental barrier readily. Some small parasites can, however, such as those found in cat droppings, and these can cause birth defects. The effects of these agents may range from inducing miscarriages to causing serious permanent defects.

Recall that viruses, which are much smaller than bacteria, can be considered similar to clusters of genes. Like genes, they are made of nucleic acid (either DNA or RNA) and have a coat of protein for protection. Because viruses normally behave by penetrating the membranes of cells, where they cause the infected cell to manufacture more virus, viruses usually have no difficulty in penetrating the membranes that are present in the placental barrier. Viruses, unlike bacteria, are resistant to antibiotics, such as penicillin and streptomycin, and are much more difficult to destroy.

German Measles (Rubella)

German measles, once considered one of the diseases children contract one time or another, was one of the first virally-induced diseases known to cause serious birth defects in women infected with the virus during the early stages of pregnancy. Physicians noticed that whenever there was an epidemic of German measles, a greater percentage of children with birth defects were born to mothers who had been infected.

Whereas infected children or adults would temporarily suffer from an incapacitating fever, numerous red rashes, and swellings behind the ear, the effects to the embryo were permanent and devastating. Children born to infected mothers could exhibit blindness, deafness, microcephaly, mental retardation, hare lip, heart defects, abnormal intestines and spina bifida. About one third of those children died before the age of six months.

The degree of damage expressed depends greatly upon the time of infection. If the mother was infected during the first, second and third month of pregnancy, the percentages of children showing serious abnormalities were fifty percent, thirty percent and seven percent respectively. The chances for damage was very low if the mother was infected during the first ten days of pregnancy or the last eight to ten weeks.

The prevalence of such defects has dropped sharply in recent years in the United States because of the widespread vaccination of children against measles. This vaccination consists of injecting the child with low doses of non-infecting *Rubella* virus. The body then builds up antibodies to the virus; these remain in the bloodstream and kill any *Rubella* viruses that might infect that individual in the future.

Likewise, after a person has contracted German measles once, usually in childhood, that person will have built up sufficient antibodies for protection. Hence, a mother who has either had the vaccine or who had been infected as a child, would not be expected to contract the disease during pregnancy. These same principles of building up antibodies from a vaccine or previous infection apply to virtually all other infectious viruses. Despite the widespread vaccinations against measles in the USA, small epidemics of measles are known to break out in regions of the country where unvaccinated new immigrants live. In some developing countries measles remains as a major killer of children.

Other Viruses

Even though measles is under control, at least in the Western world, there are still a number of other viruses known to be serious causes of birth defects. One is *cytomegalovirus* (CMV), at one time found in the urine of five percent of the women tested. The number of newborns suffering from the virus, however, is a great deal smaller. Nonetheless, CMV increases the rate of miscarriages and can lead to children with microcephaly, mental retardation, epilepsy, and cerebral palsy. Although CMV is easy to detect, there is still no good treatment for it.

In Venezuela and in the southern United States, a virus that gives horses encephalitis, can also affect the embryo and fetus of a pregnant woman. This virus will greatly increase the rate of miscarriages during the first three months of pregnancy. If the mother is infected during the fourth to the eighth months, the newborn can undergo severe damage to the brain and spinal cord.

Toxoplasmosis

This infectious agent is not a virus, but a protozoan (single-cell animal) parasite that is known to infect up to one percent of pregnant women. The parasite is normally found in many birds and animals. It is known to be present in cats and in cat droppings.

When the embryo or fetus is infected the newborns suffer similar defects as those subjected to measles or CMV. Chances of mental retardation are very high. If the mother is infected during the first trimester, there is a fifteen percent chance that the child will be infected also; during the second trimester there is a twenty-five percent chance, and a fifty percent chance during the third trimester.

Because toxoplasmosis is caused by a parasite, not a virus, it can be treated by certain drugs. Sulfadiazine is somewhat effective. The best cure is Spiramycin; however, it is not approved for use in the United States as yet. Of course it is best not to contract the parasite to begin with. To be safe, pregnant women should avoid uncooked foods and, if they have cats, should be very careful not to handle cat droppings.

Radiation

Much attention has been brought to the effects of radiation on the developing embryo and fetus by publicity on the atomic blasts at Hiroshima and Nagasaki. Studies showed that most babies born to mothers who were within a half mile of the explosion, died. Of about one hundred pregnant women within a 1.25 mile radius of the blast, twenty five had a stillborn infant, twenty five had children who died shortly thereafter, and twenty five had children who were

Table 23-1 Effects of Radiation		
Site	Effect	Time of Effect
Germ Cells	Mutation	In future generations
Somatic Cells		Varying times during lifetime
Embryo	a. No implantation	a. While zygote in oviduct
	b. Teratogenic (microcephaly, retardation, etc.) slow growth	b. Embryo, fetus, and child for life

shown to be either mentally or physically retarded, or both.

Dramatic as these results were, physicians had been aware since the early 1930s that pregnant women exposed to high doses of X-rays bore a high percentage of children with birth defects. In one study, for example, forty to fifty percent of newborns had major birth defects if their mothers had been treated with therapeutic amounts of radium or X-irradiation during pregnancy. No longer is there any doubt that radiation is harmful to the developing embryo and fetus. There is still debate, however, on the level of radiation that pregnant women can tolerate. Nonetheless, all physicians agree that great care must be taken to prevent pregnant women from being exposed to radiation, and to shield the developing child if X-irradiation becomes a medical necessity.

Damaging Effects of Radiation at Various Times and Stages of Development

Radiation (Table 23-1) can cause: (a) *developmental abnormalities* when it acts on embryonic tissues and organs; (b) *hereditary changes*, such as mutations, when it acts on genes and chromosomes of germ cells or precursor cells of the gonads; and (c) *retarded growth and cancer*, when it acts on somatic cells.

For developmental abnormalities to occur, the radiation must act on the cells of the developing tissues and organs of the early embryo. At that time the embryo is much more susceptible to the effects of radiation as compared to the cells of the mother. The teratogenic effects of radiation can be detected at birth or soon thereafter.

The *hereditary effects* of radiation, i.e. mutations within the germ cells, cannot be detected in the radiated individuals, but only in future generations of their progeny. For example, if some genes in the oocytes in the ovary of an irradiated fetus mutate, those mutations will only be detected in the future progeny of the woman developing from that fetus. The same holds true of mutations induced by radiation in the sperm and eggs of adults.

The *slowing of growth* and increase in the chance of *cancer* are both factors that can be set in motion when radiation is applied to cells of a growing individual in almost any stage of development — in the embryo, in the fetus, or in the growing child. Such changes in somatic cells might slow down the rate of division of those cells so that the organ will grow much more slowly. On the other hand, the radiation might alter the chromosomes or genes of those cells so that they may develop into cancer cells, and eventually lead to a tumor years later.

Generalizations

Clearly, the range of environmental teratogens, as well as the nature of their effects, are great. There are, however, three generalizations that hold true regardless of the teratogens being considered.

1.Factors having no noticeable effects on pregnant mothers may have great effects on their embryos.

2.The particular type of defect induced by the teratogen is related to the stage in development during which the embryo or fetus is exposed to that agent. In other words, the same teratogen can produce different defects depending upon when it is applied to the embryo or fetus.

3.Lastly, the type and degree of the birth defect is influenced by the dose of the teratogen and the overall health and genetic makeup of the mother and child.

Chapter 24

Pregnancy - Changes in the Mother

The last seven chapters, starting with Chapter 18 on fertilization, are devoted mostly to the changes occurring in the developing child as it progresses from the zygote through the blastocyst, embryonic, and fetal stages, and to the mechanisms by which the chromosomal, genetic, and environmental factors affect that development. During those nine months from fertilization to birth, the developing child acts in a manner analogous to a parasite: it depends entirely upon its host, the mother, for all of its nutrition, for space to live and grow, and for protection from the environment.

The mother, accordingly, must have: (a) an apparatus for protecting and feeding the growing embryo and fetus; (b) a means for safely expelling the newborn at the right time; and (c) the ability to feed and care for the newborn over an extended period of time. Thus, in carrying out these functions and in coping with the growing child within her, the pregnant mother undergoes significant changes in her normal (i.e. non-pregnant) body structure, physiology, and behavior. This chapter describes some of these changes. Keep in mind that each organ and system involved with pregnancy will change in different ways according to the functions that each will need to perform during pregnancy or immediately afterwards.

Changes in the Levels of Hormone

Most of the physiological changes that occur during in pregnancy are triggered by changes in the level of such hormones as estrogen, progesterone, hCG (human chorionic gonadotrophin, the pregnancy hormone) and the hormones involved in preparing the breasts to produce milk. These are hormones produced by the pituitary, ovary, placenta, and adrenal glands.

Because the body experiences major bursts in the levels of some of these hormones within a few weeks after fertilization has taken place, it is not prepared to adapt to the many physiological changes triggered by those hormones. As a result, in very early pregnancy many women experience "morning sickness," that is periods of nausea triggered by the changes caused by the new level of hormones in their bodies..

As a consequence of the increased secretion of these hormones, some of the natural endocrine glands increase in size. For example, the pituitary gland doubles in size. The thyroid gland, in order to accommodate the increased metabolism of pregnancy, enlarges. Also enlarging are the adrenal glands, which can affect the pigmentation of the skin, and the parathyroid gland, which gives off hormones affecting calcium metabolism and bone formation.

We have reviewed elsewhere in this book the various factors leading to the secretion of the ovarian (Chapt. 5) and placental (Chapt. 19) hormones, and their effects on tissues and organs.

Changes in the Uterus

The uterus performs many functions during pregnancy. It accepts the blastocyst, provides contact with the embryo and fetus by means of the placenta, protects and retains the embryo and fetus, and, finally, expels the fully developed fetus at birth. Thus, as might be expected, during the nine months of gestation the uterus changes in size and increases in weight by about two pounds, an approximately sixteen fold change. This weight increase is due mostly to the increase in size of the muscle cells of the myometrium, and somewhat to the increase in blood volume.

The uterine muscles have a number of properties and functions related to pregnancy. In the stages in which the fetus is growing rapidly, the muscles stretch greatly as they protect the fetus. To aid in expulsion of the fully developed fetus at birth, the muscles develop more at the upper end of the uterus, and not at the cervical end which must enlarge during the birth process. Likewise, the cervix itself appears to get softer during late pregnancy so that it too may expand more easily during birth for the fully grown fetus to pass through it to the vagina more easily.

Changes in the Skin

As the fetus gets larger and larger, the skin around the mother's abdomen stretches under the tension. It becomes pinkish and develops "pearly lines" ("stretch marks") there. Similar changes occur in the skin of the breast, thighs, and buttocks. The degree to which the pearly lines cover those areas, of course, varies with each woman. Sometimes those lines take a silvery white pigmentation. Other changes in the pigmentation of skin that take place in pregnant women are the so-called dark patches which can be found in the forehead and cheeks of the face, around the nipples, and along the external genitalia. These changes in pigmentation, which are induced by hormones of the pituitary and adrenal glands, usually disappear when pregnancy is over. Occasionally, however, some of the pigmentation may persist in some areas.

Other changes at the surface of the body may occur during pregnancy. For example, there may be some shedding of hair and breakage of finger and toe nails. The skin glands may become more active. Red eruptions may appear on the skin. And there may be some extra tooth decay, probably induced by the lack of calcium caused by the fetus drawing upon the mother's calcium supply in order to build fetal bones.

Changes in the Circulatory System

Because of the increased demands put on the circulatory system of pregnant women, it undergoes some major changes. First, in order to maintain an adequate supply of blood to handle the increased demands of pregnancy, the total blood volume of the mother increases by as much as six pints. This volume is extraordinarily large, sometimes doubling the normal volume of blood present in non-pregnant women. The increase in blood volume and in tissue fluids (see below) accounts for about six pounds of the weight gain occurring during pregnancy.

With this increased volume, the heart will work harder, with both the blood pressure and heart rate increasing. Likewise, as more red blood cells are manufactured, the body will need more iron to manufacture the hemoglobin present in red blood cells. The increased volume of blood and its sluggish flow often lead to an increase in the fluids of the body's tissues. When an excess of these tissue fluids accumulate and the tissues swell we have a troublesome condition known as *edema*. This condition, usually found in the legs and ankles , leads to a syndrome which sometimes accounts for the backaches and "clumsy stance" often characteristic of women in the later stages of pregnancy.

Because some of this increased volume of blood accumulates in the veins, pregnant women frequently have enlarged veins in the skin and neck, varicose veins in the legs, and hemorrhoids around the rectum.

Changes in the Breasts

Changes that occur in the breasts prepare it for the new function it will undertake when pregnancy terminates, that of providing the new infant with milk. Accordingly, during pregnancy the breasts enlarge, often accounting for an overall weight gain of about one and one half pounds. With their increase in size, the breasts may become tender to touch. The nipples and the area around them take on a darker pigmentation.

The hormones affecting the growth of the breasts, and the production and secretion of milk are described in Chapter 25. The first true milk is not produced until about the second day after birth occurs. Before that the nipples of the breast may secrete a clear proteinaceous material called *colostrum.*

Other Changes

Nausea and *vomiting* are characteristic of about sixty percent of pregnant women. In the early months the nausea is the result of the unusual increase in the hormone levels of the body and the metabolic changes thus stimulated. In the later months of pregnancy, some of the stomach discomforts are caused by the pressure of the enlarged uterus on the stomach and intestines.

This pressure, when applied to the bladder, may account for the frequent trips to the bathroom. Sometimes the bladder becomes irritated during pregnancy because of the frequent urinations, and even the process of urinating becomes uncomfortable. As might be expected, the kidney works harder during pregnancy because of its added role in eliminating the wastes of the fetus and because of the added strains on it caused by increased volume of blood and tissue fluids in the mother.

Changes are also observed in the expectant mother's behavior and emotions during pregnancy. On one hand there are the legendary weird diets with insatiable demands for pickles and chocolate ice cream sundaes in the wee hours of the morning. And because of the physiological stresses, the emotions of pregnant women may vacillate between varying degrees of happiness, anxiety, irritability, and sleepiness.

Nutritional Changes During Pregnancy

Because the nutritional demands of the growing fetus are so great, the amount and composition of the mother's diet do not remain constant throughout pregnancy. As the blood volume increases, more iron is needed. As the fetus builds more bone, more calcium is needed. Likewise the need for more protein increases, especially as the fetus grows during the last few months. If these nutrients are not supplied, not only will the growth of the fetus be slowed, but the mother's own reserves of these materials will be tapped, thereby weakening her health.

On the "Disadvantages" of Pregnancy

From the foregoing, it would appear that the process of child-bearing in humans is not very efficient. After all, the pregnant woman may suffer from nausea, vomiting, hair shedding, tooth decay, nail breaking, tender breasts, skin discoloration, constipation, hemorrhoids, varicose veins, irritated bladder, and irritability. Why do these discomforts occur?

Some believe that many of the discomforts of pregnancy stem from the upright position that evolved as humans began walking on two feet. This upright position, in turn, placed the internal positioning of the fetus and uterus in an "unnatural" relationship to the rest of the body's internal organs.

Furthermore, in order for the fetus to get through the small pelvic cavity of humans, the head can not be too large. Hence, the brain of newborn humans is relatively less fully developed than are those of most other mammals at birth. Consequently, the human newborn is more dependent upon the mother to survive than are other newborn mammals. One way that human mothers are prepared physiologically for this dependent being, the *neonate*, is by her breasts being ready to secrete milk immediately after the child is born.

Irrespective of attempts to develop biological rationalizations for the origin of these discomforts, the fact is they do exist and affect most pregnant women to some extent. With these varied aches and pains, and sometimes with the added fear of possible miscarriage, it is very tempting for pregnant mothers to take a battery of drugs including tranquilizers, and laxatives, steroids, and pain killers. In the next chapter you will learn about the harmful and permanent effects on the newborn of taking such substances. Pregnant mothers should resist these temptations with all their resolve.

What are the consolations for these discomforts? Of course that of fulfilling the ultimate creative act, the miracle of nature, that of creating a human being.

Chapter 25

Birth, Lactation, and Milk

Birth may be defined at one level as the physical processes whereby the fully developed fetus leaves the protected environment of the womb and separates from the mother to begin its life in a new environment. For most of this chapter I will focus on two groups of hormonal events within those processes: (a) those triggering the birth of the child, and (b) those controlling the production and delivery of milk to the newborn.

Birth (parturition) · Summary of the Physical Events

Pregnancy ends with the birth of the neonate. Immediately prior to birth, the mother's uterine muscles undergo a series of increasingly stronger contractions, a process known as *labor*. "Normal" birth, i.e. birth in which the neonate leaves the uterus with the crown of the head emerging first, and in which there are no complications, consists of three major stages.

First, the lower part of the uterus and cervix start to widen greatly. This widening or *dilation*, usually takes place about eight to sixteen hours before the neonate is expelled. Such, at least, is the case for a woman's first child; usually the widenings occur only four to eight hours before birth for her subsequent children.

Following this initial widening, the upper part of the uterus begins to contract. When the *contractions* become vigorous, most women experience "labor pains." These contractions usually last for about one minute, and are spaced about 10 minutes apart during the early stages of labor. After a time, when these intense uterine contractions are about 2 minutes apart, the amniotic cavity breaks and some of the amniotic fluids, the "water" is released.

In the second stage, the cervix dilates completely such that the uterus, cervix, and the vagina appear to be merged into a single broad canal. At this time the *forceful uterine contractions* expel the fetus. Usually this process is aided by the mother voluntarily "pushing" down on the uterus by taking deep breaths and contracting her abdominal region. The expelled fetus, the neonate, however, is still connected to the uterus by means of its umbilical cord and the placenta.

Finally, in the third stage about ten to fifteen minutes later, the uterus expels the placenta by means of the *decidual reaction* referred to earlier in this text. At this stage a part of the uterine wall next to the placenta is also expelled with the placenta, and the mother loses some blood. Soon afterwards, the uterus becomes smaller and firm, and occasionally contracts.

When the neonate enters the world outside of its mother, it usually arrives head first. If the buttock end comes first, we have what is known as a "breech" delivery. Because these breech deliveries are sometimes more complicated, they usually need medical assistance.

Hormonal Control of Labor

For labor to occur, the upper part of the uterus must undergo repeated contractions. Yet we have learned that the hormone progesterone inhibits the contractile activity of the uterus so that the embryo or fetus will not be expelled prematurely from the uterus. In early pregnancy, most of this progesterone comes from the corpus luteum of the ovary. In the later stages, the bulk of the progesterone that inhibits the uterus from excess contractions comes from the placenta. Thus, before labor can begin, progesterone can not be present in the uterus in large amounts. The sequence of hormonal events that lowers the level of progesterone just before labor is described below.

Role of Fetal Hormones in Initiating Labor

The normal contractions of the uterus in the absence of progesterone are not strong and frequent enough to initiate labor. Medical scientists have learned that in some mammals it is not the mother, but the fetus that starts the process of labor. It is almost as if the fetus "knows" when it is ready to enter the world; at that moment the brain of the fetus initiates a cascade of hormonal events that lead to its birth.

The chain of events is as follows (Fig.25-1): (a) the hypothalmus of the fetus stimulates its anterior pituitary to release the hormone *ACTH*. (b) The ACTH acts on the adrenal glands of the fetus, which is situated on top of the kidneys, causing them to secrete *glucocorticoid* hormones. (c) These hormones, which also serve to prepare the lungs of the fetus for breathing, cause the placenta to stop producing progesterone and to begin releasing substances known as prostaglandins. (d) The *prostaglandins* reach the uterus and cause it to start the labor contractions.

Once these initial contractions start, the mother's hormonal systems start to participate in labor. The first contractions of the uterus stimulated by the prostaglandins re-leased from the placenta, send messages to the mother's posterior pituitary gland, which, in response, begins to release the hormone *oxytocin*. This hormone stimulates the more forceful uterine contractions which eventually lead to the birth of the neonate.

Now it may seem odd that of all these hormonal events, to wit the release of ACTH, glucocorticoids, prostaglandins, and oxytocin, and the inhibition of the release of progesterone, only one, the release of oxytocin, is controlled by the mother. Did you ever wonder how scientists first discovered this amazing sequence of hormonal events?

Actually it all began with the observations of some Idaho sheep farmers. They noticed that sheep grazing on a mountain pasture where an uncommon weed grew abundantly would sometimes have extra long pregnancies ending with the deaths of the ewes and fetuses. Even more peculiar, those fetuses often were two to three times the weight of normal newborn lambs.

When scientists examined the dead fetuses, they found them to have either a missing or misplaced pituitary gland as well as underdeveloped adrenal glands. Presuming from these findings that the missing pituitary might in some way account for these long pregnancies, the scientists, by means of delicate surgery, removed the anterior pituitary of fetuses while those fetuses were still in the uterus of the ewe. The experiment showed that fetuses with their anterior pitu-

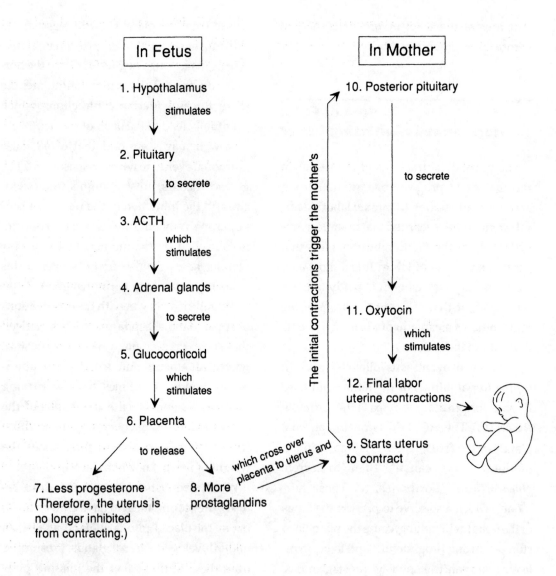

Figure 25 - 1 Hormonal control of initiation of labor.

itaries removed would also grow to two to three times their normal weight and would not be delivered on time.

The scientists concluded that the uncommon weed fed upon by the ewes contained a substance that did not hurt the ewe, but that had a teratogenic action on the developing fetus in preventing it from forming normal pituitary and adrenal glands. Hence, the original chance observation by the farmers on the sheep grazing on the strange mountain weed was instrumental in helping medical scientists understand the process by which the fetus of mammals, and probably of humans, initiates its own labor. This is but one example of how basic research and the scientific investigation of the unusual tells us much about how our bodies work.

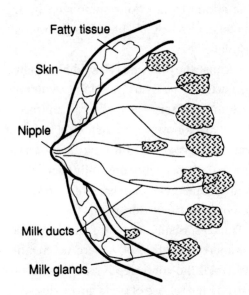

Figure 25 - 2 Internal components of breast.

Hormonal Control of Lactation

The biological role of the mother does not end with the birth of the neonate. Although she was instrumental in creating a newborn, that child is not fully developed and still is highly dependent upon the mother for a number of needs, chief among them being nutrition. Hence, the mother has to be able to manufacture milk at the right time and to develop organs and a mechanism for delivering that milk to the neonate. These functions of the mother are also primarily under the control of her hormones.

Milk Production

Even though the mature human female has relatively large breasts when not pregnant as compared to most other mammals, her breasts enlarge even more during pregnancy and remain enlarged during the months that she suckles her child. The *milk glands* (Fig. 25-2) in her breasts start to get larger beginning with the second month of pregnancy. This increase in size of the breasts during pregnancy is stimulated by the excess progesterone produced by the placenta. At the same time, however, the placental progesterone and estrogen prevent the milk (*lactogenic*) glands in the breasts from producing milk. At birth the placenta is cast off, thereby eliminating the source of most of the progesterone and estrogen as well as the inhibitory effects of those steroids on milk production (Table 25-1).

At this time the milk glands are now receptive to two particular hormones, called *lactogens* because they stimulate the production of milk. One of these lactogens is *ACTH*, a hormone which acts on the adrenal glands to influence milk production by releasing other hormones which affect the levels of sugar, fats and salts in our body fluids.

The other lactogen is *prolactin,* a hormone which directly affects the milk glands to synthesize some of the components of milk, such as the milk protein casein and the milk sugar lactose. Prolactin is the same hormone that also inhibits the ovary from ovulating.

Control of Milk Production and Ejection

When the infant suckles, it activates nerves in the nipples of the breasts to signal the hypothalmus of the mother's brain. The hypothalmus, in turn, stimulates both the anterior and posterior lobes of the pituitary gland. The anterior lobe releases prolactin, the hormone needed for manufacturing milk.

The posterior lobe, on the other hand, releases the hormone *oxytocin*, which in addition to causing the uterine contractions of labor, also activates the contractile cells of the milk glands to contract, thereby squeezing the newly formed milk into the main ducts of the nipple so that it will be ejected into the mouth of the suckling infant.

Other Actions of Oxytocin and Prolactin

Like most hormones, the oxytocin and prolactin stimulated by suckling may have other effects on the body of the mother. The oxytocin released by the suckling that occurs in the period immediately following birth also aids in the healing of the uterus. It affects this healing by increasing uterine contractions, and, by so doing, it helps get rid of bits of remaining placenta and reduces the bleeding caused by the decidual reaction. The added contractions stimulated by the oxytocin also help the uterus to return to its regular size.

Because the prolactin released by suckling also acts on the ovary to inhibit ovulation, suckling can serve as a natural contraceptive. This contraceptive action of suckling was first recognized through the observation that women who constantly suckle their young, sometimes for over eighteen months, do not menstruate during this time.

Thus, especially in poor countries where commercial contraceptives are not readily available, the contraceptive actions stimulated by the release of prolactin during long term suckling seems an effective way for spacing children every two to three years. Unfortunately, this natural contraceptive action of prolactin has been interfered with in many developing countries where the women have been encouraged to feed their

---

Table 25-1
Hormonal Control of Milk Production

I. Immediately before birth:

A. Anterior pituitary is prepared to release:
1. Prolactin to stimulate synthesis of milk components.
2. ACTH to release adrenal hormones to prepare fluids for milk.

B. Placenta releases progesterone and estrogen which:
1. Inhibit release of prolactin and ACTH.
2. Increase breast size.

II. Immediately after birth:

A. Placenta, the source of progesterone and estrogen, is removed.
B. Therefore prolactin and ACTH are ready to be released.

III. As suckling begins:

A. Anterior pituitary stimulated to release:
1. Prolactin to make milk components.
2. ACTH to release adrenal hormones to regulate water and other components of milk.

B. Posterior pituitary stimulated to release oxytocin which:
1. Causes milk ducts to contract to eject milk out of nipple.
2. Stimulates uterus to contract to get rid of last remnants of placenta.

---

infants powdered milk formulas fed by bottle. Consequently, not only did these women have children more frequently, but many of the infants died because of the lack of clean water with which to prepare the powdered milk formula.

Components of Milk

The mother does not start producing mature human milk until a day or two after the child is born. The first secretion of the breast, which can be noticed in the latter stage of pregnancy, is a clear liquid which contains more protein than mature milk and which is nutritious to the neonate. It takes about a month after the child is born until the mother's milk is completely mature and is lower in its content of protein. Once a constant regime of milk production and consumption is established, the mother will be able to produce milk continuously for periods ranging from months to over a year or two.

The mature milk of all mammals is not of the same composition. Each varies depending upon the needs of the particular species. For example, baby whales need and get a great deal of fat in their milk so that they can manufacture a layer of blubber to insulate them from the cold of the oceans. The milk from cows, for example, has over twice the protein content as does mature human milk. Hence, some dieticians suggest that if cows' milk is given to infants, that the milk be boiled first. The boiling process will make the protein of cows' milk more digestible, and will prevent the milk from curdling in the infant's stomach.

Concluding Remarks

We started this book with a detailed discussion of how hormones control the production of sperm and eggs, the receptivity of the uterus to the blastocyst, and the phases of the menstrual cycle. Following that material, very little was mentioned about hormones until Chapter 21 and the present chapter. In this chapter we have learned how hormones affect the birth process, the preparation of the breasts for nursing, and the production and ejection of milk. We also have learned about the side effects of hormones affecting those processes.

These examples serve as a reminder that the overall functioning of the body, especially those taking place from conception to birth, require the extremely fine synchronization of the neural and hormonal regulatory systems, all set in place by our genes.

The events leading to the birth of a healthy child certainly are beautiful and represent a great miracle of nature. Birth is nature's gift, and with intelligent parenting during the nine month gestation period, all should go well.

EPILOGUE

In these twenty-five chapters I have attempted to provide you with the fundamental scientific facts and principles necessary to understand the biological processes in the creation of human life. As you make plans to start your own family, this book will provide background information to help you evaluate the many news and magazine articles that appear so frequently on subjects covered in this text.

The major focus of the book is on human reproduction, human genetics, human embryonic development, and birth. It tells you about the factors necessary for everything to "go right," that is for the final result of conception to be a healthy normal child free of birth defects.

To do so the text also needed to tell you about birth defects and factors that lead to them. For example, you learned about: (1) congenital malformations caused by venereal diseases; (2) mosaics, such a hermaphrodites; (3) defects caused by chromosomal aberrations, such as Down's and Fragile X syndromes; (4) abnormalities caused by errors of fertilization, such as triploids and hydatidiform moles; (5) inherited genetic diseases, such as sickle cell anemia, galactosemia, and hemophilia; (6) abnormalities resulting from malfunctions of the sex hormone system, such as are exhibited by the gueve doces; (7) birth defects caused by such environmental factors as alcohol, radiation, and drugs; (8) those defects caused by malnutrition of the mother during pregnancy; and (9) those defects caused by hormones, such as D.E.S., taken by the pregnant mother and expressed in the child after it reaches puberty.

The text mentioned that many of the eggs which are fertilized never make it to successful birth. A large number never implant. Others implant, but abort very early, within a few weeks, without the mother even knowing that she is pregnant. And still others go through various stages of the gestation period, only to be aborted naturally through a miscarriage. Analysis of the abortuses usually show them to have some major chromosomal abnormality.

You also learned of ways to prevent severe birth defects from afflicting your offspring. For example, to avoid such chromosomal aberrancies as Down's or Turner's syndrome, females can have the cells of the early embryo analyzed for karotype by either amniocentesis or chorionic villi sampling. To avoid transmitting genetic diseases, you now know enough to construct pedigree charts for yourself and your mate, and to seek the advice of a genetic counselor. To have a healthy pregnancy, you know about the critical periods of pregnancy, about the effects of drugs, alcohol, radiation, malnutrition, and dieting.

If you understand the material in this book and apply what you have learned to modify your behavior and to take advantage of the advances of medical science during the time of your life that you will be thinking about and procreating children, then you will greatly lessen the chances that you will produce a child who has an avoidable physical or mental handicap.

The statistics indicate that from 2 to 5 percent of children born in the United States have varying degrees of physical and mental handicaps. They also show that parents who are educated have fewer children with handicaps than those with less education. It is my wish, and my major incentive for writing this book, that readers of this book will be sufficiently knowledgeable that they will have even fewer children with either congenital or hereditary birth defects.

Nonetheless, even though you think you understand the material in this book and apply its lessons to your behavior, a number of you will have a handicapped child. What if it is you? How are you going to handle it knowing that the rest of your life is going to be very different from the life that you as a college student had planned?

Will the mental strain be too much? Will you feel guilty that something that you did caused the handicap? Will you feel that you have been unfairly selected to bear this burden? Will the way you and your family handle the situation affect your other children adversely, your marriage, and lead to divorce? Will your financial circumstances give you the time to help the child, especially through the difficult early years? Will you institutionalize the child? Can you afford it? If the child does manage to survive and adjust until adulthood, how will you provide for the child's future, especially after you are dead?

Tough questions, aren't they? But they are real questions, questions that many families all over the world have to think about daily. There are no easy answers. I can only offer prospective parents of handicapped children a word of hope, hope based upon my own experiences as the father of a mentally and physically handicapped daughter.

I refer to my oldest child, Gloria, to whom I have dedicated this book. Gloria, age 34 at the time of this printing, was born prematurely with the use of forceps following the obstetrician's decision to break the amniotic sac. At first we did not believe that Gloria had any handicap, although we did suspect that everything was not just right.

I will not go into the details of Gloria's early years, but they were not easy on her and on the rest of the family. All we learned from the physicians was that Gloria was short for her age, tested as educably mentally retarded, and had a few physical defects such as a heart murmur, an unusual gait, and a partially "frozen" joint in the elbow of her right arm.

But something happened around her thirteenth birthday when she participated in her Jewish confirmation ceremony, her "Bat Mitzah." To avoid "embarrassment," we arranged to have the ceremony in a small chapel, and we hoped for the best. To the amazement of all present, as part of the ceremony, Gloria sang from memory in a beautiful soprano voice a number of ancient Hebrew melodies, including the melodious Song of Songs of King Solomon. In addition, at the reception following the ceremony, she picked up her accordion, a gift given to her just two weeks beforehand, and proceeded to play, without any instruction, a number of songs.

From that moment onwards, our lives changed. Why? Because Gloria broke us of the stereotype that mentally handicapped individuals are devoid of talent. She proved that she did have strong untapped talents, and that with a little help, she could not only enrich her own life, but the lives of most people that she touches. As a parent, I began to think more about her and less about myself. the next twenty years with Gloria have been an adventure. She has a well-trained soprano voice with perfect pitch. She plays at the highest level of accordion competition. Her repertoire of operatic and accordion pieces number over one thousand. She can converse a bit in a half dozen languages including sign language. Gloria loves to perform at schools, churches and synagogues, convalescent homes, and parties. She has appeared on numerous TV programs, talk shows, and specials. She has been featured on an award-winning PBS special, and has a fan club of students, senior citizens, government officials, and movie personalities.

The point I wish to make is that Gloria had a talent, we came to recognize it, and just as with any talented person, we needed to give Gloria the same lessons, training, and opportunities to perform that any artist needs. Although she is mentally retarded and can not carry out such simple arithmetic as adding 4 and 7, and although she has a frozen elbow joint and can not hold an accordion as others do, she can excel in music and bring great pleasure to those of us who do not have those talents.

Gloria is not unique. I have now met many talented mentally retarded citizens, some in music, some in art, some in theater, others in athletics, etc. And just as Gloria and comparable talented and gifted mentally handpicked people have been given the help to express their talents, others need the same chance.

I do not want to leave you with the conclusion that there are advantages to having children who are handicapped, and that once you find a talent in a handicapped child, the problems disappear. My main message is to treat the prospect of parenting with intelligent decisions based on knowledge, and the chances are that you will have a healthy normal child. My ancillary message is, that if for some unknown and uncontrollable reason, you do parent a handicapped child, life still can be rewarding and challenging both for you and the child. It will require love and respect for the child, and a willingness to search for that child's talents and desires, and to nurture them and provide the same help that you would provide if your child were normal and talented. You will be a better person for it.

Index

Name: _____

I.D.#: _____ Date: _____

HOMEWORK QUESTIONS FOR CHAPTER 1

■ What are the advantages and disadvantages of asexual reproduction, parthenogenesis, and sexual reproduction? Why?

_____

_____

_____

_____

_____

_____

_____

_____

_____

_____

_____

_____

_____

_____

_____

_____

_____

_____

_____

_____

Name: _____

I.D. #: _____ Date: _____

HOMEWORK QUESTIONS FOR CHAPTER 2

■ A. What is the pathway of sperm from the testis to the urethra? What glands secrete into the semen?

_____
_____
_____
_____
_____
_____
_____
_____
_____
_____
_____
_____

■ B. Describe the anatomy and physiology of an erection.

_____
_____
_____
_____

_____

_____

_____

_____

_____

_____

_____

_____

_____

_____

_____

_____

_____

_____

_____

_____

_____

_____

_____

_____

_____

_____

_____

Name: _____

I.D.#: _____ Date: _____

HOMEWORK QUESTIONS FOR CHAPTER 3

■ How do hormones regulate the production of sperm and the secondary sex characteristics?

_____
_____
_____
_____
_____
_____
_____
_____
_____
_____
_____
_____
_____
_____
_____
_____
_____
_____
_____

Name: _____

I.D.#: _____ Date: _____

HOMEWORK QUESTIONS FOR CHAPTER 4

■ Describe the formation of an egg, the fate of the follicle, and the structure of the uterus.

_____

_____

_____

_____

_____

_____

_____

_____

_____

_____

_____

_____

_____

_____

_____

_____

_____

_____

_____

Name: _____

I.D.#: _____ Date: _____

HOMEWORK QUESTIONS FOR CHAPTER 5

■ A. Describe the hormonal control of ovulation and menstruation.

_____
_____
_____
_____
_____
_____
_____
_____
_____
_____
_____
_____
_____
_____
_____
_____
_____
_____

■ B. Describe either:

        (1) The actions of female hormones after pregnancy is established.

        (2) The actions of the steroid oral contraceptives (combination type).

Name: _____

I.D.#: _____ Date:_____

HOMEWORK QUESTIONS FOR CHAPTER 6

■ Describe the physiological and hormonal basis of menstrual cramps, the failure to menstruate, and menopause.

_____

_____

_____

_____

_____

_____

_____

_____

_____

_____

_____

_____

_____

_____

_____

_____

_____

_____

_____

Name: _____

I.D.#: _____ **Date:** _____

HOMEWORK QUESTIONS FOR CHAPTER 7

■ Compare the sexual responses of females and males.

_____

_____

_____

_____

_____

_____

_____

_____

_____

_____

_____

_____

_____

_____

_____

_____

_____

_____

_____

Name: _____

I.D.#: _____ **Date:** _____

HOMEWORK QUESTIONS FOR CHAPTER 8

■ A. Compare the causes, effects, and cures of gonorrhea, syphilis, and genital herpes.

_____

_____

_____

_____

_____

_____

_____

_____

_____

_____

_____

_____

_____

_____

_____

_____

_____

_____

■ B. Describe how AIDS virus (HIV) can be spread, why it is so deadly, why finding a cure will be difficult, and the three most important rules you can follow to avoid contracting AIDS.

_____

_____

_____

_____

_____

_____

_____

_____

_____

_____

_____

_____

_____

_____

_____

_____

_____

_____

_____

_____

_____

_____

Name: _____

I.D.#: _____ **Date:** _____

HOMEWORK QUESTIONS FOR CHAPTER 9

■ A. Define a chromosome, homologous chromosomes, proteins.

_____

_____

_____

_____

_____

_____

_____

_____

_____

_____

_____

_____

_____

_____

■ B. Why do we have two sets of chromosomes in our somatic cells? Explain.

_____

_____

_____

_____

_____

_____

_____

_____

_____

_____

_____

■ C. Describe the composition, structure, and synthesis of a protein.

_____

_____

_____

_____

_____

_____

_____

_____

_____

_____

_____

_____

_____

_____

_____

_____

Name: _____

I.D.#: _____ Date:_____

HOMEWORK QUESTIONS FOR CHAPTER 10

■ A. Describe how one cell becomes two cells by mitosis.

_____
_____
_____
_____
_____
_____
_____
_____
_____
_____
_____
_____
_____
_____
_____
_____
_____
_____

■ B. Describe chromosome lag with drawings. How can this process in mitosis lead to mosaics?

_____

_____

_____

_____

_____

_____

_____

_____

_____

_____

_____

_____

_____

_____

_____

_____

_____

_____

_____

_____

_____

Name: _____

I.D.#: _____ **Date:**_____

HOMEWORK QUESTIONS FOR CHAPTER 11

■   A. What are the main differences between mitosis and meiosis? Explain.

_____

_____

_____

_____

_____

_____

_____

_____

_____

_____

_____

_____

_____

■   B. How many different arrangements of distribution of maternal and paternal
chromosomes can occur in the gametes of a species that has an n number of 10?

_____

_____

_____

_____

■ C. Describe the first and second meiotic divisions during oogenesis.

Name: _____

I.D.#: _____ **Date:** _____

HOMEWORK QUESTIONS FOR CHAPTER 12
PART I

■ A. Describe with drawings a translocation and a deletion.

_____

_____

_____

_____

_____

_____

_____

_____

_____

_____

_____

_____

_____

_____

■ B. What are the differences between chromosome lag and non-disjunction? Where can each occur?

_____

_____

_____

_____

_____

_____

_____

_____

_____

_____

_____

_____

_____

_____

_____

■ C. How can a non-disjunction lead to a monosomy or a trisomy in a zygote?

_____

_____

_____

_____

_____

_____

_____

_____

_____

_____

_____

_____

Name: _____

I.D.#: _____ Date:_____

HOMEWORK QUESTIONS FOR CHAPTER 12
PART II

■ D. Describe how it is possible to get individuals that are either XXY, XXX, or XYY. How many Barr bodies would their cells contain?

_____
_____
_____
_____
_____
_____
_____
_____
_____
_____
_____
_____
_____
_____
_____
_____
_____
_____
_____
_____
_____

_____

_____

_____

_____

_____

_____

_____

_____

■ E. What are the differences between pseudohermaphrodites and hermaphrodites?

_____

_____

_____

_____

_____

_____

_____

_____

_____

_____

_____

_____

_____

Name: _____

I.D#: _____ **Date:**_____

HOMEWORK QUESTIONS FOR CHAPTER 13

■ A. Define and give examples of the following terms:

Homozygous; heterozygous; dominant; recessive; phenotype; genotype.

_____

_____

_____

_____

_____

_____

_____

_____

_____

_____

_____

_____

_____

_____

_____

_____

_____

_____

_____

■ B. How do you determine the genotype of a pea plant with red flowers by a test cross?

_____

_____

_____

_____

_____

_____

_____

_____

_____

_____

_____

_____

_____

_____

_____

_____

_____

_____

_____

_____

_____

Name: _____

I.D.#: _____ **Date:**_____

HOMEWORK QUESTIONS FOR CHAPTER 14

■ A. What is incomplete dominance? Give an example.

_____

_____

_____

_____

_____

_____

_____

_____

_____

_____

_____

_____

_____

■ B. What is a genetic disease? Give an example of a cross between two unaffected carrier parents leading to a child showing a genetic disease (e.g. cystic fibrosis, sickle cell anemia).

_____

_____

_____

_____

_____

_____

_____

■ C. Describe in words and with a pedigree chart the genotype and phenotype of children from a marriage between a female carrier for hemophilia, and a man who does not have hemophilia.

_____

_____

_____

_____

_____

_____

_____

_____

_____

_____

_____

_____

_____

_____

_____

_____

_____

_____

Name: _____

I.D.#: _____   **Date:**_____

HOMEWORK QUESTIONS FOR CHAPTER 15

■ A. Construct a Punnet square for a cross between individuals LlGG x LlGg.

_____

_____

_____

_____

_____

_____

_____

_____

_____

_____

_____

■ B. Describe tight and loose linkage with examples. How can you test for a tightly linked gene?

_____

_____

_____

_____

_____

_____

_____

_____

_____

_____

_____

_____

_____

■ C. What blood type is a universal donor, and what blood type is a universal recipient? Give examples of the types of transfusions that are possible.

_____

_____

_____

_____

_____

_____

_____

_____

_____

_____

_____

_____

_____

_____

Name: _____

I.D.#: _____ Date:_____

HOMEWORK QUESTIONS FOR CHAPTER 16

■ A. Describe with drawings either the replication of DNA or the genetic control of the synthesis of a protein.

_____

_____

_____

_____

_____

_____

_____

_____

_____

_____

_____

_____

_____

_____

_____

_____

_____

_____

■ B. Describe either the molecular basis of mutation or the genetic regulation of the synthesis of different proteins at different times.

_____

_____

_____

_____

_____

_____

_____

_____

_____

_____

_____

_____

_____

_____

_____

_____

_____

_____

_____

_____

_____

_____

Name: _____

I.D.#: _____ Date: _____

HOMEWORK QUESTIONS FOR CHAPTER 17

■   A. Describe the sequence of events taking place during fertilization.

_____

_____

_____

_____

_____

_____

_____

_____

_____

_____

_____

_____

_____

_____

_____

_____

_____

■ B. Describe the various ways to get monozygotic twins, and the status of the placenta and amniotic cavity in each of those cases.

Name: _____

I.D.#: _____ Date: _____

HOMEWORK QUESTIONS FOR CHAPTER 18

■ Describe the formation of the neural tube and the coelom (body cavity).

_____
_____
_____
_____
_____
_____
_____
_____
_____
_____
_____
_____
_____
_____
_____
_____
_____
_____
_____

Name: _____

I.D.#: _____ Date:_____

HOMEWORK QUESTIONS FOR CHAPTER 19

■ Describe the formation of the amniotic cavity and the yolk sac.

_____

_____

_____

_____

_____

_____

_____

_____

_____

_____

_____

_____

_____

_____

_____

_____

_____

_____

_____

Name: _____

I.D.#: _____ **Date:** _____

HOMEWORK QUESTIONS FOR CHAPTER 20

■ Describe the major developmental events in the embryo from Horizon 12 through 23, and in the fetus.

_____
_____
_____
_____
_____
_____
_____
_____
_____
_____
_____
_____
_____
_____
_____
_____
_____
_____
_____

Name: _____

I.D.#: _____ Date:_____

HOMEWORK QUESTIONS FOR CHAPTER 21

■ Describe the events leading to the development of the male internal and external genitalia.

_____

_____

_____

_____

_____

_____

_____

_____

_____

_____

_____

_____

_____

_____

_____

_____

_____

_____

_____

_____

Name: _____

I.D.#: _____ **Date:** _____

HOMEWORK QUESTIONS FOR CHAPTER 22

■ Describe some of the roles that the placenta plays.

_____
_____
_____
_____
_____
_____
_____
_____
_____
_____
_____
_____
_____
_____
_____
_____
_____
_____
_____
_____
_____

Name: _____

I.D.#: _____ **Date:**_____

HOMEWORK QUESTIONS FOR CHAPTER 23

■ Compare the timing of damage to the embryo and fetus of thalidomide, measles, and radiation.

_____

_____

_____

_____

_____

_____

_____

_____

_____

_____

_____

_____

_____

_____

_____

_____

_____

_____

_____

Name: _____

I.D.#: _____ Date:_____

HOMEWORK QUESTIONS FOR CHAPTER 24

■ Summarize the changes that occur in the mother during pregnancy.

_____
_____
_____
_____
_____
_____
_____
_____
_____
_____
_____
_____
_____
_____
_____
_____
_____
_____
_____

Name: _____

I.D.#: _____ Date:_____

HOMEWORK QUESTIONS FOR CHAPTER 25

■ A. Describe the sequence of hormonal events leading to the initiation of labor.

_____

_____

_____

_____

_____

_____

_____

_____

_____

_____

_____

_____

_____

_____

_____

_____

_____

_____

_____

_____

■ B. Describe the hormonal control of milk production and ejection, and the contraceptive effects of suckling.

_____

_____

_____

_____

_____

_____

_____

_____

_____

_____

_____

_____

_____

_____

_____

_____

_____

_____

_____

_____

_____

_____

Notes

Notes